BIVARIATE DATA ANALYSIS:
A PRACTICAL GUIDE

BIVARIATE DATA ANALYSIS:
A PRACTICAL GUIDE

RANDI L. SIMS
(NOVA SOUTHEASTERN UNIVERSITY)

Nova Science Publishers, Inc.
New York

Senior Editors: Susan Boriotti and Donna Dennis
Coordinating Editor: Tatiana Shohov
Office Manager: Annette Hellinger
Graphics: Wanda Serrano
Editorial Production: Jennifer Vogt, Matthew Kozlowski and Maya Columbus
Circulation: Ave Maria Gonzalez, Luis Aviles, Raymond Davis, Melissa Diaz and Vladimir Klestov
Communications and Acquisitions: Serge P. Shohov
Marketing: Cathy DeGregory

Library of Congress Cataloging-in-Publication Data

Sims, Randi L.
    Bivariate: a practical guide / Randi L. Sims
        p. cm.
    Includes index.
       ISBN 1-56072-751-9
       1. Statistics--data processing. I. Title.

QA276.4.S56 1999
519.5—dc21

Copyright © 2000 by Nova Science Publishers, Inc.
          400 Oser Avenue, Suite 1600
          Hauppauge, New York 11788-3619
          Tel: 631-231-7269   Fax: 631-231-8175
          e-mail: Novascience@earthlink.net
          Web Site: http://www.novapublishers.com

All rights reserved. No part of this book may be reproduced, stored in a retrieval system or transmitted in any form or by any means: electronic, electrostatic, magnetic, tape, mechanical photocopying, recording or otherwise without permission from the publishers.

The authors and publisher have taken care in preparation of this book, but make no expressed or implied warranty of any kind and assume no responsibility for any errors or omissions. No liability is assumed for incidental or consequential damages in connection with or arising out of information contained in this book.

This publication is designed to provide accurate and authoritative information with regard to the subject matter covered herein. It is sold with the clear understanding that the publisher is not engaged in rendering legal or any other professional services. If legal or any other expert assistance is required, the services of a competent person should be sought. FROM A DECLARATION OF PARTICIPANTS JOINTLY ADOPTED BY A COMMITTEE OF THE AMERICAN BAR ASSOCIATION AND A COMMITTEE OF PUBLISHERS.

Printed in the United States of America

I would like to dedicate this book to my students.

Contents

List of Figures and Tables	xi
Preface	xv
Chapter 1 - Using Statistical Packages	1
Popular Statistical Packages	1
Chapter 2 - Classifying Your Data	5
Choosing the Proper Statistical Methods	5
Categorical Variables; 2 Value Limit	5
Categorical Variables; No Value Limit	5
Continuous Variables	6
Selecting the Appropriate Statistical Technique	6
Example 2-1 Steps in Testing Your Hypothesis	7
Rephrasing Your Hypotheses	8
Example 2-2 Steps in Rewriting Your Hypothesis	9
Chapter 3 - Frequency Distributions	11
Quality Control	12
Summary Statistics	13
Example 3-1 Sample Report	13
Example 3-2 Steps in Preparing Frequency Distributions	13
Cross Tabulations	13
Example 3-3 Sample Report	16
Example 3-4 Steps in Preparing Cross Tabulations	16
Chapter 4 - Descriptive Statistics	17
Measures of Central Tendency	17
Mean	18
Median	18
Mode	18

Measures of Dispersion	19
Range	19
Variance	19
Standard Deviation	19
Reporting Your Descriptive Statistics	20
Example 4-1 Steps in Preparing Descriptive Statistics	20
Example 4-2 Sample Report	21

Chapter 5 – Hypothesis Testing   23

Hypothesis versus Null Hypothesis	23
Probability	23
Support for Your Hypotheses	24
Sample Size	24
Degrees of Freedom	24
Coding	25
Continuous Variables	25
Categorical Variables	25
Reverse Scoring or Recoding	25
One- Versus Two-Tailed Tests	26
Reading Statistical Printouts	27
Preparing Tables of Your Results	

Chapter 6 - Chi-square   29

Example 6-1 Sample Report	33
Example 6-2 Sample Report	34
Example 6-3 Steps in Hypothesis Testing Using the Chi-square Statistic	35

Chapter 7 – t Tests of Two Means   37

Paired Test	38
Example 7-1 Sample Report	38
Two Independent Groups	38
Example 7-2 Sample Report	40
One-Tailed Tests	40
Paired Test	40
Example 7-3 Sample Report	41
Two Independent Groups	41
Example 7-4 Sample Report	43
Example 7-5 Steps in Hypothesis Testing Using the t Test	44

Chapter 8 – ANOVA – Analysis of Variance	**45**
Example 8-1 Sample Report	47
Example 8-2 Sample Report	48
Example 8-3 Sample Report	50
Example 8-4 Steps in Hypothesis Testing using ANOVA	50
Chapter 9 – Correlation	**51**
Positive Correlation Coefficients	51
Example 9-1 Sample Report	53
Example 9-2 Sample Report	56
Negative Correlation Coefficients	56
Example 9-3 Sample Report	58
Example 9-4 Steps in Hypothesis Testing Using Correlation	58
Chapter 10 - Testing Your Scales	**59**
Reliability	59
Test-Retest Reliability	59
Example 10-1 Sample Report	60
Internal Reliability	60
Example 10-2 Sample Report	62
Example 10-3 Sample Report	62
Validity	63
Sensitivity	63
Example 10-4 Sample Report	63
List of Hypotheses	**65**
Exercises	**67**
Selected Answers to Exercises	**83**
Glossary	**91**
Index	**95**

LIST OF FIGURES AND TABLES

Figure 5-1	Two-Tailed Bell Curve	26
Figure 5-2	One-Tailed Bell Curve	27
Figure 9-1	Scatter Plot, Age and Height	52
Figure 9-2	Scatter Plot, Diner Satisfaction and Speed of Service	55
Figure 9-3	Scatter Plot, Exercise and Stress Symptoms	57
Table 2-1	Selecting the Appropriate Statistical Technique	7
Table 3-1	Frequency Distribution – Age	11
Table 3-2	Frequency Distribution – Gender with Data Entry Errors	12
Table 3-3	Frequency Distribution – Gender	12
Table 3-4	Cross Tabulation – Age and Gender	14
Table 3-5	Cross Tabulation – Age and Gender by Working	15
Table 3A	Music Preferences, Frequency Distribution	84
Table 3B	Music Preferences by Gender, Cross Tabulation	85
Table 4-1	Selecting the Appropriate Descriptive Statistics	17
Table 4-2	Frequency Distribution – Customer Wait Time	18
Table 4-3	Frequency Distribution – Daily Sales	19
Table 4-4	Descriptive Statistics	21
Table 4-5	Frequency Distribution – Department	21
Table 4A	Long Distance Carrier, Frequency Distribution	68
Table 4B	Weight of Potato Chips in a 12 Ounce Bag, Frequency Distribution and Descriptive Statistics	69
Table 6-1	Coin Toss, Chi-square	30
Table 6-2	Coin Toss and Coin Type, Cross Tabulation and Chi-square	30
Table 6-3	Re-employment Status by Therapy Type, Cross Tabulation and Chi-square	31
Table 6-4	Re-employment Status by Therapy Type, Cross Tabulation and Chi-square, Observed (Expected Values)	32
Table 6-5	Elementary School by Testing Referral Type, Coin Toss and Coin Type, Cross Tabulation and Chi-square	34
Table 6A	Graduate Education Plans by Program, Cross Tabulation and Chi-square	72

Table 6B	Purchase Decision by Marital Status, Cross Tabulation and Chi-square	72
Table 6C	Type of Call by Day of the Week, Cross Tabulation and Chi-square	73
Table 6D	Lecture Attendance by Employee Gender, Cross Tabulation and Chi-square	73
Table 7-1	Midterm and Final Exam Scores, Paired t Test and Descriptive Statistics	38
Table 7-2	Average Employee Sick Leave Used, t Test and Descriptive Statistics	39
Table 7-3	Organizational Commitment by Leadership Training, One-Tailed, Paired t Test and Descriptive Statistics	41
Table 7-4	Average School Satisfaction by Recess Play, Two-Tailed t Test and Descriptive Statistics	42
Table 7-5	Average School Satisfaction by Recess Play, One-tailed t Test and Descriptive Statistics	43
Table 7A	Faculty and Student Views Concerning Academic Dishonesty, t Test and Descriptive Statistics	74
Table 7B	Employee Contract Term by Union Membership Status, t Test and Descriptive Statistics	75
Table 7C	Parts Returned With and Without Quality Control, One-tailed t Test and Descriptive Statistics	75
Table 8-1	Car Loan Interest Rates by Lending Institution, ANOVA and Descriptive Statistics	46
Table 8-2	Car Loan Interest Rates by Lending Institution, Lower and Upper 95% Confidence Intervals	46
Table 8-3	Grocery Purchases by Payment Type, ANOVA and Descriptive Statistics	48
Table 8-4	Average Charge for Pain Relief by Insurance Coverage, ANOVA and Descriptive Statistics	49
Table 8-5	Average Charge for Pain Relief by Insurance Coverage, Lower and Upper 95% Confidence Intervals	50
Table 8A1	Wait/Treatment Time by Day of Week, ANOVA and Descriptive Statistics	76
Table 8A2	Wait/Treatment Time by Day of Week, Lower and Upper 95% Confidence Intervals	76
Table 8B1	Time of Persistence by Student Class, ANOVA and Descriptive Statistics	76
Table 8B2	Time of Persistence by Student Class, Lower and Upper 95% Confidence Intervals	77
Table 8C1	Employee Satisfaction by Shift, ANOVA and Descriptive Statistics	77
Table 8C2	Employee Satisfaction by Shift, Lower and Upper 95% Confidence Intervals	77
Table 8D1	Book Prices by Seller, ANOVA and Descriptive Statistics	78
Table 8D2	Book Prices by Seller, Lower and Upper 95% Confidence Intervals	78

Table 9-1	Children's Age and Height, Correlation	**53**
Table 9-2	Diner Satisfaction and Speed of Service, Correlation and Descriptive Statistics	**54**
Table 9-3	Exercise and Stress Symptoms, Correlation and Descriptive Statistics	**56**
Table 9A	Employee Attitudes, Intercorrelations (Probabilities), Descriptive Statistics	**79**
Table 9A1	Employee Attitudes, Correlation and Descriptive Statistics	**89**
Table 9B	Substance Abuse Admissions, Intercorrelations (Probabilities), Descriptive Statistics	**80**
Table 9B1	Substance Abuse Admissions, Correlation and Descriptive Statistics	**89**
Table 10-1	Job Satisfaction, Internal Reliability	**61**
Table 10A	Intrapersonal Role Conflict, Test-retest Reliability	**80**
Table 10B	Continuance Commitment, Internal Reliability	**81**

PREFACE

The purpose of this book is to help the users of computerized statistical packages make the correct statistical choices to match the data they have already collected. It is assumed that the reader has information available on the use of the statistical package. While most statistical packages will perform whatever statistical analyses the user chooses, the package will not help the user in the selection of the choice or in the interpretation of the results. Thus, a statistical package is only a tool. While statistical packages come with instructions on their use, the user must know how to apply the package to the data available and how to read the results printed.

This book will provide a simple to understand, step-by-step guide for the user of any computerized statistical package. This book is designed for bivariate data analysis. Bivariate means that two variables are tested together.

Chapter 1

USING STATISTICAL PACKAGES

For bivariate hypothesis testing, just about any student version, commercial statistical software, or mainframe package will meet your needs. Oftentimes, the decision to choose one package over another has to do with prior experience or access. If you have used a statistical package in the past (as a class requirement, for example), continuing to use the same package will give you the benefit of prior experience. Unless you have become an expert user, expect to rely on the user's manual when running the program. The user's manual will provide you with the steps and codes to use for each statistical test you select. The remainder of this book will help you in the selection of these statistical tests and provide guidelines on reading the printed results.

POPULAR STATISTICAL PACKAGES

The following summarizes some of the Statistical Packages commonly used by social science researchers. Mainframe packages (typically in UNIX format) have the ability to handle huge data sets quickly. Student versions tend to provide fewer statistical techniques than the desktop packages, but their prices are significantly reduced. Comprehensive packages offer more statistical tests than the average statistical user will ever need. These packages are geared toward scientific research. Universities typically provide mainframe statistical packages for their faculty and students.

MATLAB® - MATRIX LABORATORY

MATLAB is targeted towards engineers and scientists. It provides extensive mathematical computations and advanced graphics in addition to statistical testing. MATLAB is available in a mainframe, desktop, and student package. MATLAB is a registered trademark of The MathWorks. www.mathworks.com

MINITAB – NOT AN ACRONYM

Minitab may be one of the most popular software programs introduced to students. Minitab offers an extensive list of statistical tests. It is available in a desktop (windows) or student package. www.minitab.com

NCSS – NUMBER CRUNCHER STATISTICAL SOFTWARE

NCSS advertises that it specializes in "providing statistical analysis software to the occasional user of statistics." Over 200 statistical procedures are offered. NCSS is available in a desktop (windows) package. www.ncss.com

PASS – POWER ANALYSIS AND SAMPLE SIZE

Offered by the same organization that offers NCSS. PASS advertises that it is "easy to learn, easy to use accurate, complete, and easy to interpret." Over 20 statistical tests are offered. PASS is available in a desktop (windows) package. www.ncss.com

SAS – NOT AN ACRONYM

Formally known as Statistical Analysis System. SAS is a comprehensive statistical program. It is available in a Unix and desktop package. www.sas.com

SPSS – STATISTICAL PRODUCT AND SERVICE SOLUTIONS

Formally known as Statistical Package for the Social Sciences (SPSS). SPSS is a comprehensive statistical program. It is available in a Unix, desktop (Windows and Macintosh), and student package. www.spss.com

STATA – STATISTICS FOR DATA ANALYSIS

STATA is a comprehensive statistical program. It is available in a Windows, DOS, Macintosh, and Unix format. www.stata.com

STATISTICA – NOT AN ACRONYM

STATISTICA is a comprehensive statistical program offered by StatSoft. It is available in a Windows, Macintosh, and student version. www.statsoftinc.com

STATISTIX - Statistical Analysis Software for Personal and Professional Use

Statistix is a comprehensive statistical program for Windows. It is advertised as "extraordinarily easy to learn and use." www.sigma-research.com/bookshelf/rtsxw.htm

SYSTAT – System for Statistics

SYSTAT is a comprehensive statistical and graphics program for scientists, engineers, and statisticians. It is available in a Windows, DOS, Macintosh, and student package. www.spss.com

VISTA – The Visual Statistics Program

VISTA is advertised as designed for teachers and students. It is available "freely redistributable" from its web site. This program offers dozens of statistical tests for univariate and bivariate needs. It is available in a Windows, Macintosh, and Unix format. forrest.psych.unc.edu/research/ViSta.html

Chapter 2

CLASSIFYING YOUR DATA

CHOOSING THE PROPER STATISTICAL METHODS

Choosing the proper statistical methods depends upon the classification of your variables. This chapter is useful for hypotheses or objectives which specify the relationship between two variables (called bivariate). If your hypothesis is worded to include more than two variables, you might want to rephrase that hypothesis into more than one. See the section in this chapter on rephrasing your hypothesis for examples.

Before you can select the appropriate statistical techniques for your hypothesis, you must classify each of your variables into one of three classifications; categorical variables with a limit of 2 values, categorical variables with no limit on the number of values, or continuous variables. An explanation of the three classifications follows.

CATEGORICAL VARIABLES; 2 VALUE LIMIT

A categorical variable simply divides the values of the variable into independent groups. An example of a categorical variable with only two values is Gender. One of the values is male and the other value is female. Other examples might be Shift (day, night), Location (east, west), Answer (yes, no), or Group (experimental, control). Any variable with a limit of only two values would fulfill this classification.

CATEGORICAL VARIABLES; NO VALUE LIMIT

An example of a categorical variable with more than two values is Color. You might classify Color into many groups. For example, red, yellow, green, blue, and pink are all possible values for the variable Color. Other examples might be Shift (first, second, third), Department (accounting, marketing, customer service), Location (east, west, south, north), or Group (experimental 1, experimental 2, experimental 3, experimental 4, control). As you see, some of the same variables are used in examples here that were mentioned earlier. The classification you use depends upon how you measured your variable. If you measured your variable with only two values, you would use the previous classification. If you measured your variable with more than two values, you would use this classification.

CONTINUOUS VARIABLES

Continuous variables are different from categorical variables in that the continuous variables exist between their measures. Adults tend to measure their ages in years. But, as a continuous variable, Age can also be measured as fractions of years. Other examples of continuos variables are Education (number of years), Weight, Height, Sales, Absences, Answers Correct, and most attitude scales. Note that we might use a category to measure these variables instead; Education (graduated, did not graduate), Absence (present, absent), etc. In these later cases, those variables would be classified as categorical and not continuous. Again, the classification depends upon how you measured your variable.

Attitude scales tend to be Likert-type measures (named after Rensis Likert who designed this type of measure). Likert-type measures provide a range of responses like: strongly agree, agree, neither agree nor disagree, disagree, strongly disagree. Other attitude scales might use responses like: very satisfied, satisfied, neither satisfied nor dissatisfied, dissatisfied, and very dissatisfied. All scales of this type are coded with numerical values ranging from low to high (1 to 5) or high to low (5 to 1). Attitude scales often have five values, but if there are more than five choices in the attitude scale used, the researcher would assign more than five values. While there might be a mathematical debate on whether these types of scales are simply categories and not continuous variables, social scientists do agree that they can safely be used in statistical analyses as continuous variables.

SELECTING THE APPROPRIATE STATISTICAL TECHNIQUE

Once you have classified your two variables into one of these three categories, you then transfer those classifications to Table 2-1. Table 2-1 works similar to mileage charts found on maps or multiplication tables. As with either of these types of charts/tables, order does not matter (i.e. the mileage from New York to San Diego is equal to the mileage from San Diego to New York and 8 x 2 = 2 x 8).

Most hypotheses state that something is related to or influenced by something else. To begin, you must recognize the variables, then you must classify each variable. Using Table 2-1 as a guide, select the statistical test that is appropriate for your data. See Example 2-1 for step-by-step directions.

Table 2-1
Selecting the Appropriate Statistical Technique

	Categorical 2 value limit	Categorical no value limit	Continuous
Categorical 2 value limit	chi-square	chi-square	*t* test
Categorical no value limit	chi-square	chi-square	ANOVA
Continuous	*t* test	ANOVA	correlation

After you determine the appropriate statistical test, you can turn to the chapter describing that technique for an explanation. Alternately, you can just choose that technique from your computerized statistical package. Each chapter provides help and examples for understanding the output of the computerized statistical package.

Example 2-1
Steps in Testing Your Hypothesis

Step 1: Write your hypothesis.

Hypothesis 2.1 - Employee location is related to level of satisfaction.

Step 2: Identify your variables.

The first variable is Location and the second variable is Satisfaction.

Step 3: Classify your variables.

How is Location measured?

A. As one of two different places? If so, then the variable Location would be a categorical variable with two values.
B. As one of three or four (or more) locations? Then the variable Location would be a categorical variable with no limit on values.
C. As distance in miles from the corporate office? If so, that would be a continuous variable.

How is Satisfaction measured?

A. As one of two choices, yes or no? If so, then the variable Satisfaction would be a categorical variable with two values.

B. As one of more than two choices, yes, no, or can't decide? Then the variable Satisfaction would be a categorical variable with no limit on values.
C. As a Likert-type scale ranging from very satisfied to very dissatisfied? If so, that would be a continuous variable.

Step 4: Select your statistical test using Table 2-1.

Once you have classified both of your variables, Table 2-1 will show you the appropriate statistical analysis to test your hypothesis. In this example of Hypothesis 2.1, Location is one of two different places (a categorical variable with only two values) and Satisfaction is the sum of a Likert-type scale (a continuous variable). Using Table 2-1, the t test is the appropriate statistical test for this hypothesis.

Step 5: Turn to the chapter on the appropriate statistical test.

REPHRASING YOUR HYPOTHESES

A hypothesis with two variables is a bivariate hypothesis (bivariate means two variables). Bivariate hypotheses can be tested using bivariate statistical tests. A hypothesis with more than two variables is a multivariate hypothesis. Multivariate hypotheses must be tested with multivariate statistical tests. This book only focuses on bivariate statistical testing.

If your hypothesis includes more than two variables you may want to rephrase it into two independent hypotheses. Most likely you are interested in the independent influences for each variable and not simultaneous interactions. If you have one hypothesis with more than two variables and you test each pair independently, your data may only partially support that one hypothesis. With two separate hypotheses, one hypothesis can be supported and the other rejected. See Example 2-2 for one hypothesis that has been rewritten as two.

Example 2-2
Steps in Rewriting Your Hypothesis
One Hypothesis With Three Variables

Step 1: Write your hypothesis.

Hypothesis 2.2: Learning time is related to age and gender.

Step 2: Identify your variables.

The first variable is Learning Time, the second variable is Age, and the third variable is Gender.

Step 3: Write two hypotheses, each with only two variables.

Hypothesis 2.2a: Learning time is related to age.

Hypothesis 2.2b: Learning time is related to gender.

Step 4: Classify your variables.

Learning Time and Age are continuous variables and Gender is a categorical variable with only two values.

Step 5: Select your statistical test using Table 2-1.

Using Table 2-1, Hypothesis 2.2a would be tested with a correlation since both Learning Time and Age are continuous variables. Hypothesis 2.2b would be tested with a t test since Gender is a categorical variable with only two values and Learning Time is a continuous variable.

Step 6: Turn to the chapter on the appropriate statistical test.

Note:

If your statistical testing found that Learning Time was related to Age but not to Gender, Hypothesis 2.2 could neither be accepted or rejected. Yet, using the divided hypotheses, Hypothesis 2.2a would be supported and Hypothesis 2.2b would be rejected.

Chapter 3

FREQUENCY DISTRIBUTIONS

A frequency distribution arranges your data in a table-like display according to the values recorded for each variable. The display will provide a listing of each value recorded for the variable, the number of times (frequency) that value was recorded, and the percentage of time a value was recorded. Most computer generated frequency distributions will also provide cumulative information. Cumulative information (appropriate for continuous variables) is simply a summation of those values that came before. See Table 3-1 for an example of a frequency distribution.

Table 3-1
Frequency Distribution - Age

Value	Frequency	Cumulative Frequency	Percentage	Cumulative Percentage
18	3	3	6%	6%
19	17	20	40%	46%
20	4	24	8%	52%
21	4	28	8%	56%
22	7	35	14%	70%
24	1	36	2%	72%
28	4	40	8%	80%
29	1	41	2%	82%
33	3	44	6%	88%
41	1	45	2%	90%
42	2	47	4%	94%
45	1	48	2%	96%
51	1	49	2%	98%
66	1	50	2%	100%

There are two main reasons for preparing frequency distributions for your data. The first reason is a check of quality control for the data entry process. The second reason is as summary statistics. Computer generated frequency distributions should be prepared for all variables in every study.

QUALITY CONTROL

It is quite likely that there will be at least one data processing error when transferring the survey data into the computerized data base. Since fields of numbers are hard to read on a computer screen, determining where an error has occurred may be difficult. The frequency distribution is one method of checking on the accuracy of your data entry. For example, the variable Gender should only have two possible values. Selecting 1 for males and 2 for females is a common coding method (any two numbers are equally appropriate). When preparing a frequency distribution, only those two numbers should be displayed. If any other number is listed on the frequency table, you would know that a data entry error has occurred. See Table 3-2 for an example of a computer generated frequency distribution with errors. Note that if a data entry error has occurred, but the value still falls within the appropriate range of values (a 1 instead of a 2 for Gender), the frequency distribution will not be useful for locating this error.

Table 3-2
Frequency Distribution - Gender with Data Entry Errors

Value	Frequency	Cumulative Frequency	Percentage	Cumulative Percentage
1	23	23	46%	46%
2	25	48	50%	96%
4	1	49	2%	98%
8	1	50	2%	100%

After preparing a frequency distribution for each variable, you should carefully evaluate each print-out to ensure accurate data entry. If any errors are found, corrections should be made in the data set and new frequency distributions prepared before proceeding any further. See Table 3-3 for a corrected frequency distribution for Gender.

Table 3-3
Corrected Frequency Distribution - Gender

Value	Frequency	Cumulative Frequency	Percentage	Cumulative Percentage
1	24	24	48%	48%
2	26	50	52%	100%

SUMMARY STATISTICS

Frequency distributions make excellent summaries of your data. The tables are useful for reporting purposes, are a necessary step in descriptive statistics (see Chapter 4), and are a simple way to display your data. See Example 3-1 for a sample report based on Table 3-1 and Table 3-3.

Example 3-1
Sample Report

The sample included 24 (48%) males and 26 (52%) females (see Table 3-3). The ages of the students varied widely, from a low of 18 to a high of 66 (see Table 3-1). Although this is a wide range, 70% of the students were between the ages of 18 and 22, with the most frequent age 19 years, representing 40% of the students.

For a step-by-step guide for preparing frequency distributions, see Example 3-2.

Example 3-2
Steps in Preparing Frequency Distributions

Step 1: Prepare a frequency distribution for each variable in your study.

Step 2: Check each frequency distribution for accuracy.

Step 3: Correct errors detected.

Step 4: Prepare new frequency distributions as needed.

Step 5: Use the frequency distributions to prepare summary reports.

CROSS TABULATIONS

A specialized way to prepare the frequency listings of two or three variables simultaneously is called a cross tabulation. A cross tabulation provides a great deal of information about the relationship between variables. To statistically test this relationship, see Chapter 5 on chi-square. You might be interested in the frequency of the variable Age divided by the variable Gender. Table 3-4 displays a cross tabulation for two variables. Instead of a frequency listing by Age only or Gender only, the cross tabulation displays both variables simultaneously. For example, not only do we show the number of 17 year olds (18 students), but we also show how many of them are male (7 students) and how many are female (11 students).

Table 3-4
Cross Tabulation - Age and Gender

Age	Gender		Total
	Male	Female	
17	7	11	18
18	11	9	20
19	12	13	25
20	16	14	30
21	2	5	7
Total	48	52	100

In addition to displaying two variables in a cross tabulation, we can also display three variables. This display is slightly different from the one table display for two variables. Multiple tables are included in the display, at least one table for each value of the third variable. To display the variables Age, Gender, and Working, we must prepare a table of Age and Gender for each value of Working. Since Working, in this example, is measured as one of three values, (1) part-time employment, (2) full-time employment, and (3) not working at all, we must prepare a three-part table.

Table 3-5 displays a cross tabulation for these three variables. The first table displays only those students employed part-time and breaks down their characteristics by Age and Gender. The second table displays only those students employed full-time and breaks down their characteristics by Age and Gender. The third table displays only those students not employed at all and breaks down their characteristics by Age and Gender. So, by looking at these tables, for example, we could show how many students are working part-time and are 17 year old females (7 students). See Example 3-3 for a sample report based on Table 3-5.

Table 3-5
Cross Tabulation - Age and Gender by Working

Part-time Employment

Age	Gender		Total
	Male	Female	
17	3	7	10
18	9	6	15
19	4	5	9
20	6	4	10
21	2	0	2
Total	23	22	45

Full-time Employment

Age	Gender		Total
	Male	Female	
17	2	2	4
18	1	2	3
19	4	4	8
20	8	9	17
21	1	4	5
Total	16	21	37

Not Working At All

Age	Gender		Total
	Male	Female	
17	2	2	4
18	1	1	2
19	4	4	8
20	2	1	3
21	0	1	1
Total	9	9	18

> **Example 3-3**
> **Sample Report**
>
> Of the 100 students sampled, 37 reported that they were employed full-time. Of these, 59% reported that they were 20 years old or older (9 males and 13 females). Younger students (male and female alike) reported less full-time employment then did older students. Only 18% of 17 and 18 year olds reported full time employment, while 46% of 20 and 21 year olds did (see Table 3-5).

Typically cross tabulations are only prepared when the variables have a limited number of values. It would provide little additional information, for example, if your values for the variable Age ranged from 2 to 97, or if there were 10 values for Working. For a step-by-step guide for preparing cross tabulations, see Example 3-4.

> **Example 3-4**
> **Steps in Preparing Cross Tabulations**
>
> Step 1: Select two or three variables that together describe your data well. Multiple pairs or sets may be selected.
>
> Step 2: Check the frequency distributions to ensure a limited number of values for each variable in your cross tabulation list.
>
> Step 3: Prepare cross tabulations.
>
> Step 4: Use the cross tabulations to prepare summary reports.
>
> *Note*: Cross tabulations should not be used in place of hypothesis testing.

Chapter 4

DESCRIPTIVE STATISTICS

Statistics which describe your data fall into two general categories. The first category describes the central tendency of your data. There are three common measures of central tendency: mean, median, and mode. The second category describes the dispersion of your data. There are three common measures of dispersion: range, standard deviation, and variance. The appropriate descriptive statistics should be prepared for all variables in every study. Table 4-1 can be used as a guide for determining when each descriptive statistics is appropriate. Refer to Chapter 2 for information concerning the classification of your variables.

Table 4-1
Selecting the Appropriate Descriptive Statistic

	Categorical Variables	Continuous Variables
Mean		✔
Median		✔
Mode	✔	✔
Range		✔
Standard Deviation		✔
Variance		✔

MEASURES OF CENTRAL TENDENCY

Measures of central tendency describe how the data fall together around the middle (or central) numbers. Preparing a frequency distribution (Chapter 3) demonstrates how the numbers might cluster around the middle. Describing how the data is grouped (or not grouped) is useful for summarizing your data. A frequency distribution is shown in Table 4-2 so that measures of central tendency can be prepared for the variable Customer Wait Time.

Table 4-2

Frequency Distribution - Customer Wait Time (in minutes)

Value	Frequency	Cumulative Frequency	Percentage	Cumulative Percentage
1	5	5	5%	5%
2	4	9	4%	9%
3	5	14	5%	14%
4	12	26	12%	26%
5	9	35	9%	35%
6	16	51	16%	51%
7	27	78	27%	78%
8	7	85	7%	85%
9	6	91	6%	91%
10	3	94	3%	94%
11	6	100	6%	100%

Mean. This descriptive statistic is calculated as an arithmetic average. The mean is an appropriate statistic for continuous variables only. Calculating the mean is never appropriate for categorical variables. The mean can not be determined by the frequency distribution alone. Using a computerized statistical package, the mean for the variable Customer Wait Time has been calculated as 6.12 minutes.

Median. The median is the middle number when all values for a variable are listed from lowest to highest. Duplicate values are listed however many times they were recorded. The median is an appropriate statistic for continuous variables. The median can be determined by reading the frequency distribution prepared for any continuous variable. Using Table 4-2, the median Customer Wait Time is determined by reading the value represented by the 50th cumulative percentage. In this case, the 50th percentage value is found within the set of 6 minutes wait time. The value of 6 minutes wait time includes the 36th percentage through the 51st percentage. The reason that the value of 6 minutes wait time spans such a large section of percentages is that 6 minutes wait time has been recorded 16 times (the frequency). The first time the value of 6 minutes was recorded, the cumulative percentage was 36, the second time the value of 6 minutes was recorded, the cumulative percentage was 37, etc. Note that the value of 7 minutes includes the 52nd through the 78th percentage.

Mode. The mode is the most frequently occurring value for any variable. The mode is an appropriate statistic for all variables. Using Table 4-2, the mode for the variable Customer Wait Time is 7 minutes. This means that most customers (27 out of the 100) within this sample, waited 7 minutes for service.

Thus, looked at together, the three measures of central tendency for the variable Customer Wait Time describe that the data are grouped around 6 and 7 minutes wait time. The median is 6, the mode is 7, and the mean is 6.12. Each of these descriptive statistics of central tendency accurately describe the variable Customer Wait Time. The similarity of the three measures is not always the case. When differences are found, care must be taken to ensure that the reported descriptive statistics are a good summary of your data. Accuracy of the statistics is not enough, the appropriate statistics must also be reported.

Table 4-3 displays a frequency distribution for the variable Daily Sales. As can be seen, there are large differences between the three measures of central tendency. The median is $900, the mode is $600, and the mean is $1018. While all three statistics are mathematically correct, the mean does a poor job of describing the data. Both the median and mode represent the central tendency of the variable Daily Sales much better than the mean does.

Table 4-3
Frequency Distribution - Daily Sales

Value (dollars)	Frequency	Cumulative Frequency	Percentage	Cumulative Percentage
600	21	21	42%	42%
900	10	31	20%	62%
1000	10	41	20%	82%
1200	2	43	4%	86%
1500	3	46	6%	92%
2000	3	49	6%	98%
6400	1	50	2%	100%

MEASURES OF DISPERSION

Measures of dispersion describe how much variability there is within the values selected for the variables. Measures of dispersion are only calculated for continuous variables.

Range. The range describes the values of the variable from the lowest to the highest. The range is an appropriate statistic for continuous variables. Using Table 4-2, the range of Customer Wait Time is from 1 to 11 minutes. Using Table 4-3, the range of Daily Sales is $600 to $6400. Thus, we can conclude that there is a greater dispersion for the variable Daily Sales.

Variance. The variance is an appropriate measure of dispersion for continuous variables. It is calculated as the average squared difference each value is from the mean. The larger the number, the greater the dispersion. The variance can not be calculated directly from the frequency distribution. Using a computerized statistical package, the variance for Customer Wait Time is 6.03 minutes and the variance for Daily Sales is $7435.61. This measure of dispersion also demonstrates that the variable Daily Sales is much more disperse than the variable of Customer Wait Time. Since the definition of the variance is the sum of the squared difference each value is from the mean, it is difficult to relate this measure directly back to the data collected. The data points are not squared values, as is the variance. To make interpretation easier, the standard deviation is often reported instead of the variance.

Standard Deviation. The standard deviation is an appropriate measure of dispersion for continuous variables. It is calculated as the square root of the variance. Using a computerized statistical package, the standard deviation for Customer Wait Time is 2.45 minutes and the standard deviation for Daily Sales is $86.23. The standard deviation is often reported instead of the variance since it is not a squared measure. Again, this measure of dispersion demonstrates that the variable Daily Sales is much more disperse than the variable of Customer Wait Time.

REPORTING YOUR DESCRIPTIVE STATISTICS

Once the appropriate descriptive statistics have been prepared for all of your variables, you will use those statistics to describe your data to the reader. In addition to a report in paragraph form, appropriate tables should be prepared to display the results of your descriptive statistics. The sample size should also be included within the tables of descriptive statistics. For a step-by-step guideline for preparing descriptive statistics, see Example 4-1.

Example 4-1
Steps in Preparing Descriptive Statistics

Step 1: Prepare means and standard deviations for all continuous variables.

Step 2: If means do not represent the central tendency of the data well, prepare median and mode statistics.

Step 3: Write the appropriate report of your descriptive statistics.

Step 4: Prepare frequency reports for your categorical variables.

Table 4-4 is an example of a table displaying descriptive statistics for the continuous variables Age, Absence, and Satisfaction. Frequency distributions are typically all that are needed for displaying categorical variables. Table 4-5 is an example of a frequency distribution for the variable department which might be prepared for summarizing your categorical data. Note that Table 4-5 does not report cumulative information, which would be inappropriate for categorical variables. Computerized statistical packages do not differentiate between categorical and continuous variables. Thus, frequency tables generated by statistical packages will display cumulative values for all variables. It is up to the user of these packages to determine which information should be included in a report of the data. See Example 4-2 for a sample report based upon Table 4-4 and Table 4-5. Note that more information may be given in the tables than actually reported in paragraph form.

Table 4-4
Descriptive Statistics

Variable	Sample Size	Mean	Standard Deviation	Range
Age	61	32.52	7.64	20 to 49
Absence	66	8.45	5.10	3 to 21
Satisfaction	66	75.33	12.49	42 to 96

Table 4-5
Frequency Distribution - Department

Department	Frequency	Percentage*
Accounting	15	22.7
Laboratory	17	25.8
Maternity Care	10	15.2
Pharmacy	10	15.2
Physical Therapy	14	21.2

*May not sum to 100% because of rounding; sample size = 66.

Example 4-2
Sample Report

Sixty six employees participated in the survey, representing 5 hospital departments (see Table 4-5). The average reported age for the employees in the sample was 32.52 years (5 employees did not report their age) with a standard deviation of 7.64 years. Number of days absent per year ranged from a low of 3 to a high of 21. The average reported days absent was 8.45. On a possible scale from a low of 20 to a high of 100, the average job satisfaction was 75.33 with a standard deviation of 12.49. Descriptive statistics are shown in Table 4-4.

Chapter 5

HYPOTHESIS TESTING

By now you have written your hypotheses and selected the appropriate statistical techniques to test them. Before you read the statistical results from these statistical tests, there are a few things you should know.

HYPOTHESIS VERSUS NULL-HYPOTHESIS

Researchers do not typically state their null-hypothesis in their prepared reports. This does not mean that the null-hypothesis does not exist, however. Null-hypotheses are always worded to state that there are no differences or no relationships. Your alternate hypothesis, or hypothesis for short, states that there are differences between the groups or relationships among the variables. Your hypothesis should always be written in your report, the null-hypothesis should be included only if specifically requested.

PROBABILITY

The probability level is a positive number between 0.0000 and 1.00. A probability of zero means that what you tested is not at all likely to be found unless real differences (or relationships, etc.) actually exist. A probability of 1.00 means that what you tested is quite likely to be found without any real differences (or relationships, etc.). Probability is the likelihood that no differences exist in the data you have tested. It does not matter if you have selected a t test, a chi-square, a correlation, or ANOVA. Every statistical test of your hypotheses will provide a probability. The probability is sometimes called the significance level. The computerized software might print the probability as p, p value, prob, or sig. Some software programs use capital letters, others do not. The probability level should be included in the report of your results.

SUPPORT FOR YOUR HYPOTHESES

To determine if you should accept or reject your hypothesis, you must compare the probability level to an established cut-off. Most social scientists use the 0.05 cut-off. Infrequently, the 0.01 cut-off is used. Unless specially requested to use a different cut-off, you should use the cut-off of 0.05. This means that the probability level printed with your statistical test must be less than or equal to 0.05 before you can accept your hypothesis. If the printed probability is greater than 0.05, your hypothesis must be rejected. Do not report that your results are "close to acceptance". You must either accept or reject. You should not round the probability that has been printed by the software to determine acceptance. For example, when $p = 0.050002$, you must reject your hypothesis. Sometimes researchers do not specify their established cut-off. When not stated, it can be assumed to be 0.05. Do not use the word "prove" or state that your hypothesis is "true". Unfortunately, due to the limitations inherent in social science research, we are never able to prove anything to be true. The best we can do is to accept or support our hypotheses.

SAMPLE SIZE

The sample size is the number of people or items in the group you have studied. If you surveyed 30 children in one classroom, your sample size is 30. Sample size is often printed as n by statistical software. If you gathered 3 items for each student, but only 28 answered item number 2, then you would have 2 *missing values*. Software typically utilizes as many cases as are complete for each statistic you have requested. In this example, the tests that utilized items 1 and 3 would have $n = 30$; but for items 1 and 2, $n = 28$. The two cases which had values for item 1 but not for item 2 would be excluded in this second test. Some statistical tests use n in calculating the probability level. Those tests which do not use the sample size, use the degrees of freedom instead. The sample size should be included in your reporting of results.

DEGREES OF FREEDOM

The degrees of freedom is related to the sample size or the number of groups your data has been divided into. Degrees of freedom will be printed as df or $d.f.$ by the software. Each statistical test which utilizes df has a formula for calculating the degrees of freedom. In its simplest form, $df = n - 1$. Since the df may utilize another formula (depending on the statistical test), you should not assume that $df = n - 1$. The degrees of freedom should be included in the reports of your results when printed by the software.

CODING

The information you have collected must be coded for use with statistical software packages. Coding means that you assign a numerical value for each reply. The numerical value may be the actual answer, like 25 for age, or may be a number you assign, like 1 for males and 2 for females. Variables can be coded as continuous variables or as categorical variables.

Continuous Variables. Most continuous variables will retain the value the subject provided. If an employee writes that he/she has been employed for 12 years, you will use 12 as the coded value. Similarly, a value of 8 years recorded by an employee will be coded as 08. Be careful to standardize the number of digits needed for each variable value. Sometimes people use fractions of years. If this is the case, you must decide, in advance, how you will handle this type of reply. You can convert all years into months, use decimals, or you could round up or down to the next whole year. The choice is yours, as long as you are consistent and state in your report your response to such fractions. Survey items which measure attitudes often have a number pre-assigned to the reply. For example, it is common to assign 5 to *strongly agree*, 4 to *agree*, 3 to *neither agree nor disagree*, 2 to *disagree*, and 1 to *strongly disagree*. Thus, higher points are an indication of stronger agreement. Instead, you may choose to have higher points represent stronger disagreement. Your hypothesis testing will not be affected by the coding you select.

Categorical Variables. Categorical variables must be assigned codes by you. These codes are quite arbitrary, the numbers will simply specify the group. For example, it is common practice to code males as 1 and females as 2. But you might choose any two numbers. Variables like department, location, job title, treatment, etc. must all be assigned codes for each value selected by the subjects.

REVERSE SCORE OR RECODING

Some items in your survey may be positively worded and others may be negatively worded. If agreement with either leads to higher scores, then the scaling must be reverse scored for one of these items. I might ask for your level of agreement for a series of items from 1 *strongly disagree* to 5 *strongly agree*. If my first item asks "My supervisor explains requests clearly", agreement with this item can be interpreted positively. Yet, agreement with "My supervisor often looses patience with employees" is interpreted negatively. If greater scores are to be interpreted as positive feedback concerning supervisors, than the second item must be reverse scored prior to statistical analysis. This means that instead of assigning 5 points for *strongly agree*, 1 point is assigned. Instead of 1 *strongly disagree* to 5 *strongly agree*, this type of item will be coded as 5 *strongly disagree* to 1 *strongly agree*. Oftentimes the software program can handle reverse scoring for particular survey items with a simple command statement. If this is not possible, then these particular items must be recoded by hand during data entry.

ONE- VERSUS TWO-TAILED TESTS

One- and two-tailed tests refer to the normal distribution or bell-shaped curve. With a two-tailed test, using the 0.05 cut-off, we are allowing for 0.025 percent for each tail of the bell curve. With a one-tailed test, we are allowing for the entire 0.05 percent in one tail. Figure 5-1 displays a two-tailed display of the 0.05 cut-off in a bell curve. Figure 5-2 displays a one-tailed display of the 0.05 cut-off in a bell curve.

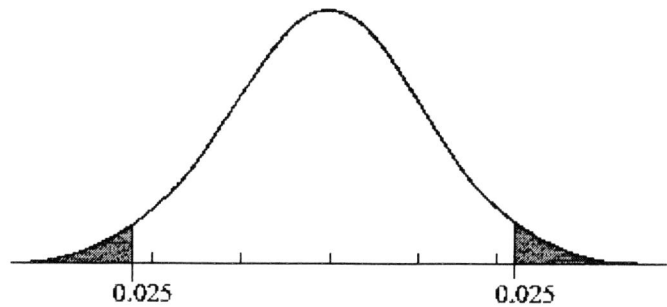

Figure 5-1
Two-Tailed Bell Curve

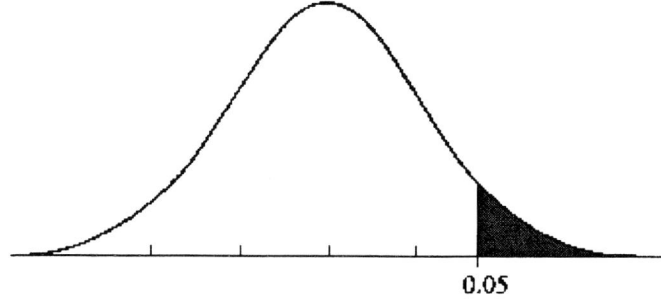

Figure 5-2
One-Tailed Bell Curve

If your hypothesis is written with a specific direction of differences, a one-tailed test should be selected. For example, "Group 1 has higher scores than group 2." This is different than proposing that "Scores for group 1 are different than scores for group 2." In the first example, the hypothesis proposes a specific direction of the difference. This example should be tested with a one-tailed test. The second example does not specify the direction of the differences. This second example should be tested with a two-tailed test. Occasionally the software program will not offer the choice between a one-tailed or a two-tailed test. If a one-tailed test is not available, the two-tailed test should be run and the probability level adjusted before comparing the result to the established cut-off of 0.05. The computed probability for a one-tailed test is ½ of the probability of a two-tailed test. If the $p = 0.0624$ for a two-tailed test, then $p = 0.0312$ for a one-tailed test.

READING STATISTICAL PRINTOUTS

When you read the printed results you should think about what each statistic means. If the results indicate that the average age for your sample is –24.5 or 135.2, you should know that this is not possible. It is easy to make data entry errors, and just as easy to copy incorrect information from the statistical printouts. The results should make sense to you. If they do not, look for data entry errors or recoding errors. The frequency distribution (see Chapter 3) is a good place to begin your search for errors if your results are out of the expected range.

While each software program may print the results of the testing you have selected slightly differently from another program, they should all have some basic components. As each statistical test is introduced in the following chapters, the commonly printed results are described.

PREPARING TABLES OF YOUR RESULTS

Statistical software packages often provide information to the user that is not typically included in your report. Do not include the statistical tables printed verbatim in your final paper. Instead, you can follow the guidelines suggested in the chapters that follow for each statistical test.

Chapter 6

CHI-SQUARE

The chi-square statistic is appropriate for categorical variables with no limit on the number of values. Refer to Chapter 2 for information concerning the classification of your variables. Chi-square may be printed as χ^2 in statistical print-outs.

The chi-square statistic utilizes the cross-tabulation table introduced in Chapter 3. The computer generated analysis for chi-square provides the cross-tabulation table, along with the chi-square value (χ^2), degrees of freedom (*df*), and probability (*p*).

Chi-square tests whether what you observed is different from what you expected to observe. Chi-square always expects that the outcome will be fair or even. For example, if you toss a coin ten times, you would expect that the coin would land on five heads and five tails. Chi-square expects the same thing. If you toss the coin 11 times, you know it is impossible to observe 5 ½ heads and 5 ½ tails, but chi-square would expect this anyway. What you observe must be significantly different from the expected. As discussed in Chapter 5, the test of significance (also called the probability level) is typically 0.05 in social science research. So, even if you observed 5 heads and 6 tails when you flipped the coin 11 times, chi-square would indicate that your observations (5 and 6) are not significantly different from the expected (5 ½ and 5 ½).

HYPOTHESIS 6.1 – COIN TOSS IS UNFAIR

The results obtained using a computerized statistics package are displayed in Table 6-1. To accept the hypothesis that differences do exist, the probability or significance level must be compared to the established cut-off of less than or equal to 0.05 (see Chapter 5). Since the probability is greater than the 0.05 cut off, we can see that what we observed is not significantly different from what we expected to observe. Our hypothesis: "Coin Toss is unfair", is therefore rejected.

Table 6-1
Coin Toss
Chi-square

	Coin Toss		Total
	heads	tails	
	5	6	11

$\chi^2 = 0.0909$; $df = 1$; $p = 0.7630$.

In the above example about the coin toss, there was only one variable. This variable had two possible values, heads and tails. Your hypothesis will contain two variables (each of them categorical). The coin toss example can be expanded to include a second variable. Let us say that you tossed a quarter, a dime, and a nickel. Now this example has two variables, Coin Toss and Coin Type. Coin Toss still has two values (heads and tails) and Coin Type has three values (quarter, dime, nickel). Table 6-2 displays the results of 20 tosses for each type of coin.

HYPOTHESIS 6.2 – COIN TOSS BY COIN TYPE IS UNFAIR

As shown in Table 6-2, the results of the Coin Toss **appear** different from what both we and chi-square would expect. After all, we know that we should have 10 heads and 10 tails for each Coin Type. The only way to know if these differences are significant, is for us to evaluate the chi-square statistic and probability. Reading our probability of 0.0839; we can see that it is greater than our cut-off of 0.05. Thus, our hypothesis of "Coin toss by coin type is unfair" is rejected. While the table may look different from what we might expect, the results are not significantly different.

Table 6-2
Coin Toss and Coin Type
Cross Tabulation and Chi-square

Coin Type	Coin Toss		Total
	heads	tails	
quarter	7	13	20
dime	11	9	20
nickel	14	6	20
Total	32	28	60

$\chi^2 = 4.96$; $df = 2$; $p = 0.0839$.

Coin tossing provides an easy example for understanding the differences between observations and expectations. However, it is unlikely that anyone would ever actually collect data on tossing coins. More likely you are interested in differences between groups of people or treatments. Instead, let us consider the effects of different types of physical therapy treatments on the employment status of patients. Patients could have received one of three types of treatments, home therapy, clinic-based therapy, or no therapy. Re-employment status will be measured three months after the injury as full-time, part-time, or not working at all.

HYPOTHESES 6.3 - RE-EMPLOYMENT STATUS IS RELATED TO THE TYPE OF THERAPY RECEIVED

Since both re-employment status and type of therapy are categorical variables (see Chapter 2 for more information on classifying your variables), the chi-square statistic is the appropriate test for this hypothesis. The computerized statistics program provides a cross-tabulation table, the chi-square statistic, degrees of freedom, and probability. Table 6-3 displays this information. The chi-square test expects that there will be no differences for re-employment status based on the type of treatment.

To accept the hypothesis that differences do exist, the probability or significance level must be compared to the established cut-off of less than or equal to 0.05 (see Chapter 5). The probability level for the chi-square statistic generated with this data is 0.0016. This probability is less than 0.05, so the hypothesis is accepted. The data demonstrates that there is a difference in re-employment status dependent upon the type of therapy provided.

Table 6-3
*Re-employment Status by Therapy Type
Cross Tabulation and Chi-square*

Re-employment	Therapy			Total
	home	clinic	none	
full-time	8	21	6	35
part-time	13	5	5	23
not at all	2	0	4	6
Total	23	26	15	64

$\chi^2 = 17.48; df = 4; p = 0.0016$

After you have accepted your hypothesis, you should then look at the data to see where the differences were uncovered. Careful attention should be paid to the differences between the observed and expected patient counts for each cell (box on the table). Expected patient counts should not be displayed in your reports, but they can be useful in determining where the observed data differed from the expected. Table 6-4 displays a reproduction of the computerized print-out, which includes the expected values for each cell in parentheses.

Table 6-4
Re-employment Status by Therapy Type
Cross Tabulation and Chi-square
Observed (Expected) Values

Re-employment	Therapy			Total
	home	clinic	none	
full-time	8 (12.58)	21 (14.22)	6 (8.20)	35
part-time	13 (8.27)	5 (9.34)	5 (5.39)	23
not at all	2 (2.16)	0 (2.44)	4 (1.41)	6
Total	23	26	15	64

$\chi^2 = 17.48$; $df = 4$; $p = 0.0016$

Far more clinic therapy patients returned to work full-time (21) than were expected (14.22). Yet the home and no therapy groups (8, 6) returned to full-time employment less frequently than expected (12.58, 8.20). In addition, we expected that 2.44 patients in the clinic therapy group would still not be working, when in fact, no clinic therapy patients had failed to return to work. The home therapy group reported that they were not working at all as was expected (2 versus 2.16) yet the no therapy group were much more likely to remain not working (4) than was expected (1.41). Note that differences in part-time re-employment are more difficult to interpret on face value. After all, the only reason why the clinic therapy group were less likely to be working part-time (5 versus the expected 9.34), is because they were working full-time.

If your hypothesis was rejected because the probability was greater than 0.05, there is no need to compare where observed values were different than expected. This is because there were no **significant** differences. Do not describe **apparent** differences when the statistical test indicates that there are no significant differences. Example 6-1 provides a sample report based on these results.

> Example 6-1
> Sample Report
>
> Seventy patients were contacted by telephone three months after their initial injury. Of these 70 patients, 6 were previously retired from employment and were not included in the subsequent analysis. The remaining patients were asked to report their re-employment status as working full-time, part-time, or not working at all. Patients were also asked about the type of physical therapy they received for their injuries. Replies were coded as home therapy, clinic-based therapy, or no therapy at all.
>
> The results of the chi-square statistic indicate that significantly ($\chi^2 = 17.48$; $p = 0.0016$) more patients receiving clinic-based therapy had returned to full-time employment than would be expected if therapy treatments were unrelated to re-employment status (see Table 6-3).

As another example, let us consider the differences between schools for the number of children who are referred to psychologists for testing in intelligence (IQ), attention deficit disorder (ADD), and hyperactivity (H).

HYPOTHESIS 6.4 – TESTING REFERRALS DIFFER BY ELEMENTARY SCHOOL

In Hypothesis 6.4, there are two variables. One variable is testing referral. It is categorical with the three values mentioned earlier (IQ, ADD, H). The second variable is school. It is also categorical and includes the number of schools in the sample. In this case, we will consider five elementary schools (E1 to E5). Since the schools vary in the number of students enrolled, the methodology includes a random observation of records for 200 children from each school. If there are no referral differences between the five schools, we would expect (as would chi-square) a similar number of referrals for each school for each of the three testing procedures. Table 6-5 displays the results of the chi-square performed by the computerized statistics program.

Table 6-5
*Elementary School by Testing Referral Type
Cross Tabulation and Chi-square*

School	Testing Referral*			Total
	IQ	ADD	H	
E1	5	3	7	15
E2	8	5	6	19
E3	4	5	11	20
E4	8	10	7	25
E5	3	4	3	10
Total	28	27	34	89

*IQ = intelligence; ADD = attention deficit disorder; H = hyperactivity. $\chi^2 = 6.274$; $df = 8$; $p = 0.6166$

The data indicate that the number of referrals for psychological testing between these five elementary school vary from a low of 10 for school E5 to a high of 25 for school E4. To determine if these differences are significant, we must compare the printed probability of the chi-square to the established cut-off of 0.05. Our results indicate that the probability of these differences occurring, if no true differences actually existed, is 0.6166. This means that 61% of the time we will find differences as large as the ones found, even though there are no actual differences in the schools for referrals made for psychological testing. While this range may appear to indicate differences between the five schools, the results of the chi-square indicate that there are no significant differences between these schools. Thus, our hypothesis of "Testing referrals differ by elementary school" is rejected. Example 6-2 provides a sample report based on these results.

Example 6-2
Sample Report

Five elementary schools were randomly selected from those schools within the city boundaries. From each of these five schools, 200 student records were randomly selected for analysis. Students must have attended the school for at least one entire academic year to qualify for inclusion in the study. Student records were reviewed to determine if the student had been referred for psychological testing. Testing referrals were coded as intelligence (IQ), attention deficit disorder (ADD), and hyperactivity (H).

A chi-square test was conducted to determine if there were any differences between the number of referrals made by school. The results of the chi-square statistic indicate that there were no significant ($\chi^2 = 6.274$; $p = 0.6166$) differences between the schools for referrals made for psychological testing (see Table 6-5).

For a step-by-step guideline for hypothesis testing using the chi-square statistic, see Example 6-3.

Example 6-3
Steps in Hypothesis Testing Using the Chi-square Statistic

Step 1: Double check that both of the variables in your hypothesis are categorical (see Chapter 2).

Step 2: Run the chi-square statistic using your computerized statistical program. If not printed by default, ask for the cross tabulation tables.

Step 3: Compare the calculated probability (also called significance level) to the standard cut-off of 0.05 (see Chapter 5).

Step 4: If the calculated probability is greater than the 0.05 cut-off, reject your hypothesis and write the appropriate report of no significant differences found.

Step 5: If the calculated probability is less than or equal to the 0.05 cut-off, accept your hypothesis. Study the expected and observed cell differences in the cross tabulation table to determine where the differences were uncovered. Write the appropriate report for your significant differences.

Chapter 7

t TESTS OF TWO MEANS

The *t* test is appropriate for hypotheses which have one categorical variable dividing the sample into two groups (two value limit) and one continuous variable. This may take the form of a paired test or two independent groups. The computer generated analysis for the *t* test provides the mean for each group ($\bar{x}$), the standard deviation (*SD*), the *t* value (*t*), the degrees of freedom (*df*), and the probability or significance level (*p*).

PAIRED TEST

The paired *t* test compares the score each subject received on his/her first test to the score of the second test. Since this is paired, the exact score from each subject for one measurement is compared to the score from the second measurement the subject received. This may be as simple as 20 people measuring their weight on two different bathroom scales. The two weights for each person are compared to determine if there are significant differences. To be significant differences, the probability or significance level obtained with the *t* test must be compared with the established cut-off of 0.05 (see Chapter 5). It is not sufficient that the two weights appear different, they must be significantly different. The paired *t* test is often used in the test-retest design.

Some students believe that they perform better on the second test an instructor gives because they become familiar with how the instructor tests. Other students feel that final exams are more difficult than midterm exams. Let us consider whether the midterm and final exam scores for students in a particular class are different.

HYPOTHESIS 7.1 – MIDTERM AND FINAL EXAM SCORES FOR STUDENTS IN SECTION 102 ARE SIGNIFICANTLY DIFFERENT FROM EACH OTHER

Hypothesis 7.1 has two variables. One variable is Exam Type. It is a categorical variable with two values, midterm and final. The second variable is Exam Score. It is a continuous variable (see Chapter 2 for help in the classification of your variables). Since we have two scores for each student, the paired *t* test is the appropriate test of this hypothesis. Table 7-1 displays the computer generated results of the paired *t* test.

Table 7-1
Midterm and Final Exam Scores
Paired t Test and Descriptive Statistics

	Midterm	Final
Mean	85.73	89.00
Standard Deviation	8.58	6.47
Sample Size	33	33

$t = -1.72; df = 32; p = 0.0952$

As we can see, the average midterm exam score is 85.73 and the average final exam score is 89.00. These scores appear different. The final exam score appears higher than the midterm exam score. In fact, the negative sign on the *t* value (-1.72) is an indication that the second score in the pair is greater than the first. However, the probability (0.0952) is greater than the cut off of 0.05, so the hypothesis of "Midterm and final exam scores for students in section 102 are significantly different from each other" must be rejected. The differences we see are simply normal variability which might be found in this size sample of scores.

While the students' grades have been paired for the hypothesis testing, typically the paired data is not displayed in the table provided the reader. When requested, raw data may be included in the appendix of your manuscript. See Example 7-1 for a sample report based on these results.

Example 7-1
Sample Report

Midterm and final exam scores were gathered for the 33 students enrolled in section 102. Using a paired t test, the results indicate that the students' scores on the midterm exams were not significantly different from their scores on the final exams (t = -1.72; p = 0.0952), see Table 7-1.

TWO INDEPENDENT GROUPS

When the research design does not call for two measures for the same people, but one measure for two different groups of people, the *t* test for two independent groups is used. The *t* test for two independent groups measures whether the scores from one group are significantly different from the scores for the other group. Instead of specific scores being compared, the average score for one group is compared with the average score for the second group. There are times when information has been collected twice from the same group, but the researcher is unable to match the tests. This may be because of anonymity issues or if there were subjects absent during one or the other collection time. In these types of cases, the *t* test for independent groups may be used.

A company might be interested in whether a change in sick leave policy is related to the average number of sick leave days taken by its employees. Sick leave has been measured by the human resources department for years, so the data is available for comparison. Last year (year 1) the sick leave policy stated that unused sick leave would be paid to employees at the end of December. This year (year 2) the sick leave policy states that unused sick leave will be accrued indefinitely for future needs. The change in policy corresponded with a change in long term disability benefits, and was not expected to influence the use of sick leave at all. Thus, the hypothesis does not state an expectation that one year will be higher than the other. If one year was hypothesized to be higher than the other, a one-tailed (see chapter 5) *t* test would be used.

HYPOTHESIS 7.2 – AVERAGE EMPLOYEE SICK LEAVE USED IN YEAR 1 IS DIFFERENT FROM THE AVERAGE EMPLOYEE SICK LEAVE USED IN YEAR 2

Hypothesis 7.2 has two variables. One variable is Year. It is a categorical variable with two values, year 1 and year 2. The second variable is Sick Leave. It is a continuous variable measured in days. Since we have two averages, one for each year, the *t* test is the appropriate test of this hypothesis. Table 7-2 displays the computer generated results of the *t* test.

Table 7-2
Average Employee Sick Leave Used (in days)
t Test and Descriptive Statistics

	Year	
	1	2
Mean	1.979	1.728
Standard Deviation	0.44	0.35
Sample Size	103	108
$t = 4.59$; $df = 209$; $p = 0.0000$		

As shown in Table 7-2, the average sick leave used in year 1 ($\bar{x} = 1.979$) seems similar to the average sick leave used in year 2 ($\bar{x} = 1.728$). Apparent similarities may be deceiving. You must always check your probability level and compare it to the established cut-off of 0.05. The results here indicate that the probability of 0.0000 is less than the cut-off of 0.05, thus the hypothesis of "Average employee sick leave used in year 1 is different from the average employee sick leave used in year 2" is accepted. We may then look back at the results to determine where the differences have occurred. From the positive *t* value (*t* = 4.59) we know that the second measurement is less than the first measurement. Or, we can simply compare the two averages and see that significantly fewer sick days were used in year 2 than were used in year 1. Example 7-2 provides a report based on these results. The report includes information not included in Table 7-2 (the total number of sick days used), but this information can be calculated using the information provided in Table 7-2 (the mean times the number of employees for the year).

Example 7-2
Sample Report

In year 1, 103 employees used 204 sick days ($\bar{x}$ = 1.979) and in year 2, 108 employees used 187 sick days ($\bar{x}$ = 1.728). The average employee sick leave used in year 1 was compared to the average employee sick leave used in year 2 using a *t* test of two means. The results indicate that employees used significantly (t = 4.59; p = 0.0000) fewer sick leave days in year 2 than they used in year 1, see Table 7-2.

ONE-TAILED TESTS

Both Hypothesis 7.1 and 7.2 were written without a specific direction of differences in mind. This is not always the case, however. Sometimes researchers hypothesize that the one measure or group, for example, will be greater than the other. If the hypothesis is written as such, than a one-tailed *t* test must be used. One-tailed tests can be performed for data gathered from paired or independent groups.

Paired Test. The top managers of a hotel have recently measured their employees' level of commitment to the organization. Much to their displeasure, commitment levels were found to be significantly lower than industry averages. After studying the survey results, managers theorize that the low levels of commitment may be related to poor supervisor-employee relations. In an attempt to improve employee commitment levels, all employees with supervisory responsibilities within one property were enrolled in a comprehensive leadership training program. Six months after training, employees were again asked to complete the organizational commitment survey.

HYPOTHESIS 7.3 – HOTEL EMPLOYEE ORGANIZATIONAL COMMITMENT WILL INCREASE FOLLOWING SUPERVISORY LEADERSHIP TRAINING

Hypothesis 7.3 has two variables. One variable is Organizational Commitment. It is a continuous variable measured as the number of points on an attitude scale. The second variable is Supervisory Leadership Training. It is a categorical variable measured as before training and after training. Since we have a pair of Organizational Commitment scores for each employee, one before and one after training, the paired *t* test is the appropriate test of this hypothesis. Since our hypothesis has been written with a specific direction in mind (increase), a one-tailed test must be used. Table 7.3 displays the results of the one-tailed, paired *t* test for this data.

Table 7-3
Organizational Commitment by Leadership Training
One-tailed, Paired t Test and Descriptive Statistics

	Organizational Commitment	
	before training	after training
Mean	42.03	63.25
Standard Deviation	11.65	8.80
Sample Size	105	105

$t = -13.92$; $df = 104$; $p = 0.0000$

As we can see in Table 7-3, the average employee organizational commitment score before leadership training was 42.03 and the average employee organizational commitment score after leadership training was 63.25. Thus, employee organizational commitment increased by over 21 points following leadership training. However, before we can accept or reject our hypothesis, we must consider if this increase is a significant increase. After comparing our probability of 0.0000 to the established cut-off of 0.05, we can accept our hypothesis of "Hotel employee organizational commitment will increase following supervisory leadership training." Example 7-3 provides a sample report based on this data.

Example 7-3
Sample Report

One-hundred and five hotel employees from one property were surveyed to determine their level of organizational commitment prior to and six months following supervisory leadership training. The average employee organizational commitment before leadership training was 42.03. Following training, employee organizational commitment increased by over 21 points for an average of 63.25. A one-tailed, paired t test was used to determine if employee organizational commitment increased significantly following leadership training. The results indicate that the employees expressed significantly greater organizational commitment following supervisory leadership training (t = -13.92; p = 0.0000), see Table 7-3.

Two Independent Groups. Based On The Related Literature Reviewed, The School Counselor Proposes That Children Will Express Greater Satisfaction With School If Provided Recess On A Daily Basis.

HYPOTHESIS 7.4 – CHILDREN WHO PARTICIPATE IN DAILY RECESS PLAY WILL EXPRESS GREATER SCHOOL SATISFACTION THAN CHILDREN WHO DO NOT PARTICIPATE IN RECESS PLAY

Hypothesis 7.4 has two variables. One variable is Recess Play. It is categorical with two values (daily play, no play). The second variable is School Satisfaction. It is continuous, measured in number of satisfaction points expressed. Since we have two satisfaction averages, one for each group of children, the t test is the appropriate test of this hypothesis. In Hypothesis 7.4, we can see that a specific direction of differences is proposed. We are proposing that the daily recess play children will be more satisfied than the no recess play children. Some computerized statistics programs perform only two-tailed t tests. The user must make a simple correction to the printed probability if the hypothesis is written for a one-tailed t test. As discussed in Chapter 5, the two-tailed probability level must be divided by 2 to obtain the one-tailed probability level. Table 7-4 displays the computer generated results of the two-tailed t test. Since we hypothesized a one-tailed t test, but the computer printed a two-tailed probability, we must adjust the expressed probability level. In our earlier example (Hypothesis 7.3, Table 7.3) we hypothesized a one-tailed test and the computer generated a one-tailed probability. No adjustment was necessary in this case. Adjustment is only necessary when the software does not print the one-tailed probability.

Table 7-4
Average School Satisfaction by Recess Play
Two-tailed t Test and Descriptive Statistics

	Recess Play	
	daily	none
Mean	25.37	19.28
Standard Deviation	11.92	14.48
Sample Size	35	38

$t = 1.95; df = 71; p = 0.0549$

As shown in Table 7-4, the average school satisfaction for children receiving daily recess play ($\bar{x} = 25.37$) appears greater than the average school satisfaction for children receiving no recess play ($\bar{x} = 19.28$). However, the printed probability level of 0.0549 is greater than the established cut off of 0.05. If our hypothesis had stated that there was a difference between the two groups, but we did not propose the direction of the difference, then we would reject this hypothesis. Since we hypothesized that the play children would be more satisfied than the no play children, we use the adjusted probability of 0.02745 (0.0549 ÷ 2). Table 7-5 displays the results of the one-tailed t test with the corrected probability. Given that the adjusted probability is less than the cut-off of 0.05, our hypothesis of "Children who participate in daily recess play will express greater school satisfaction than children who do not participate in recess play" is accepted. Example 7-4 provides a report based on the results of the one-tailed t test.

Table 7-5
*Average School Satisfaction by Recess Play
One-tailed t Test and Descriptive Statistics*

	Recess Play	
	daily	none
Mean	25.37	19.28
Standard Deviation	11.92	14.48
Sample Size	35	38

$t = 1.95; df = 71; p = 0.0274$

Example 7-4
Sample Report

Random assignment was used to place two sections of fourth grade students into the daily recess play group (n = 35) and two sections of fourth grade students into the no recess play group (n = 38). A one-tailed t test was used to determine if the daily recess play children expressed significantly greater school satisfaction ($\bar{x} = 25.37$) than the no recess play children ($\bar{x} = 19.28$). The results indicate that the children assigned to the daily play group expressed significantly greater school satisfaction than did the children assigned to the no play group (t = 1.95; p = 0.0274), see Table 7-5.

One-tailed tests and their associated tables should always be labeled as such, or the reader will assume a two-tailed test. A one-tailed test must be hypothesized prior to the selection of the appropriate statistical test. There must be a theoretical reason for the selection of a one-tailed hypothesis which has been sufficiently justified. It is inappropriate to rewrite a two-tailed hypothesis that would otherwise be rejected. For a step-by-step guideline for hypothesis testing using the *t* test, see Example 7-5.

Example 7-5
Steps in Hypothesis Testing Using the *t* Test

Step 1: Double check that one of your variables divides your data into two groups (categorical, with a two value limit) and that your second variable is continuous (see Chapter 2).

Step 2: Decide if you should use the paired *t* test, for data collected on the same group at two separate times, or the *t* test for two independent groups.

Step 3: Decide if you are using a one-tailed or a two-tailed test. Your hypothesis will be written as general differences for a two-tailed test and as a specific difference (i.e. one group is higher than the other) for a one-tailed test.

Step 4: Run the appropriate *t* test using your computerized statistical program.

Step 5: Compare the calculated probability (also called significance level) to the standard cut-off of 0.05. If using a one-tailed test, adjust the probability level if a one-tailed probability level has not been printed by the software (see Chapter 5).

Step 6: If the calculated probability is greater than the 0.05 cut-off, reject your hypothesis and write the appropriate report of no significant differences found.

Step 7: If the calculated probability is less than or equal to the 0.05 cut-off, accept your hypothesis. Study the two group means to determine where the differences were uncovered. Write the appropriate report for your significant differences.

Chapter 8

ANOVA – ANALYSIS OF VARIANCE

In Chapter 7 we considered testing the differences between two groups. In this chapter, we expand that coverage to more than two groups. Analysis of Variance, or ANOVA, is appropriate for hypotheses which have one categorical variable dividing the sample into three or more groups and one continuous variable. Analysis of Variance can also be used in multivariate hypothesis testing. When only two variables are used in ANOVA, it is called one-way analysis of variance. This chapter will cover the one-way analysis of variance. The computer generated analysis for ANOVA provides the mean for each group ($\bar{x}$), the standard deviation (*SD*), the *F*-ratio (*F*), the degrees of freedom (*df*), and the probability or significance level (*p*).

Like the *t* test, ANOVA compares the average responses for the different groups in your study. Unlike the *t* test, there is not a limit on the number of groups ANOVA can compare. In this next example, we will compare the average yearly interest charged for a new car loan by four different lending institutions over a six month period.

HYPOTHESIS 8.1 – INTEREST CHARGES FOR NEW CAR LOANS DIFFER BY LENDING INSTITUTION

We have two variables in Hypothesis 8.1. The first variable is Interest Charges. It is a continuous variable measured in percentage points. The second variable is Lending Institution. It is a categorical variable measured as one of four different institutions. One-way analysis of variance is the appropriate statistic to test this hypothesis. The yearly interest charge gathered weekly over a six month period will be compared for each of the four different lending institutions. Table 8-1 displays the computer generated results of the ANOVA.

Table 8-1

Car Loan Interest Rates by Lending Institution
ANOVA and Descriptive Statistics

	Lender 1	Lender 2	Lender 3	Lender 4
Mean	11.98	11.97	11.94	11.64
Standard Deviation	0.3514	0.3586	0.3569	0.3666
Sample Size (in weeks)	26	26	25	26

$F = 5.26; df = 3; p = 0.0022$

The sample size for three of the four lenders is 26 weeks. Lender 3 has a sample size of 25 weeks because the interest rate for one week was not reported, thus there was one missing value. The average new car loan interest rate for the four lending institutions ranges from a high of 11.98% to a low of 11.64%. These rates appear quite similar to each other. Before we can accept or reject Hypothesis 8.1, we must compare the probability reported to the standard cut off of 0.05 (see Chapter 5). The probability of 0.0022 is less than 0.05, so we can accept our hypothesis of "Interest charges for new car loans differ by lending institution".

The significant probability tells us that these four lenders do not charge the same rate. What we do not yet know, however, is which of these four lenders is significantly different from the others. For example, all four could be significantly different from the others, two might be different from the other two, or just one might be different from the other three. To determine where the significant differences have been found, subsequent tests can be performed. Choices differ by the software used, but typically tests such as Bonferroni, Scheffe, and Tukey are useful for determining where the group differences are significant. Often times follow-up testing is automatically printed by the software immediately following the initial results. If given the choice, choose "All" for statistics requested. Comparisons of the printed lower and upper confidence intervals for each group mean can also help to determine which groups differ. Table 8-2 displays the calculated lower and upper 95% confidence intervals provided by the statistics software. A table displaying the lower and upper confidence intervals is not typically prepared for the reader.

Table 8-2

Car Loan Interest Rates by Lending Institution
Lower and Upper 95% Confidence Intervals

	Lender 1	Lender 2	Lender 3	Lender 4
Lower	11.84	11.82	11.79	11.49
Upper	12.12	12.11	12.09	11.78

The lower and upper intervals establish a range of interest rates that estimates (using the standard error of the mean) the true population mean for the lender. The 95% confidence interval is a standard in the social science field. In studying Table 8-2, we can see that the lower and upper intervals for lenders 1, 2, and 3 overlap each other. Thus, there are no significant differences in yearly interest rates for these three lenders. The intervals printed for lender 4, however, do not overlap the other three lenders. Thus, follow-up testing indicates that lender 4 charges significantly lower rates for new car loans than the other three lenders. A sample report based on this analysis of variance can be found in Example 8-1.

Example 8-1
Sample Report

The yearly rates for new car loans charged by four lending institutions were recorded weekly over a six month period. The average loan rates for each of the four lenders were compared using analysis of variance. The average loan rates ranged from a high of 11.98% to a low of 11.64%. The results of the ANOVA indicate that lender 4 charged significantly lower rates ($F = 5.26$; $p = 0.0022$) for new car loans than did the other three institutions, see Table 8-1.

For a second example, let us consider the relationship between payment method and purchases made at a local grocery store.

HYPOTHESIS 8.2 – THERE IS A RELATIONSHIP BETWEEN PAYMENT METHOD AND PURCHASES MADE

Hypothesis 8.2 has two variables. One variable is Payment Method. It is a categorical variable with three values (cash, check, or credit/debit card). The second variable is Purchases. It is a continuous variable measured in dollars. The one-way analysis of variance is the appropriate test of Hypothesis 8.2. We have three groups of customers. One group pays for their groceries by cash, a second group pays by check, and the third group pays with a credit or debit card. The data will be collected for one day from a single full-service register. Table 8-2 displays the computer generated results of the ANOVA.

Table 8-3
Grocery Purchases by Payment Type
ANOVA and Descriptive Statistics

	Cash	Check	Credit/Debit
Mean (in dollars)	67.80	71.00	66.81
Standard Deviation	13.80	12.00	13.67
Sample Size	69	30	36

$F = 0.88; df = 2; p = 0.4166$

The results indicate that 69 customers paid for their purchases by cash, 30 paid by check, and 36 paid with a credit or debit card. The average purchases ranged from $66.81 for the debit/credit group to $72.00 for the check group. The probability level of 0.4166 is greater than the cut-off of 0.05. Thus, the hypothesis of "There is a relationship between payment method and purchases made" must be rejected. There is no need for follow-up testing or a comparison of confidence intervals. See Example 8-2 for a sample report of these results.

Example 8-2
Sample Report

A total of 135 customers were serviced by the test register. The average purchase made for these 135 customers was $68.24, see Table 8-3. The results of the ANOVA indicate that there is not a significant relationship between the type of payment method used and the amount of purchase made ($F = 0.88; p = 0.4166$).

As a third example, let's consider the average charges associated with pain-relief for patients admitted to a single hospital over the last year. Patients will be grouped by insurance coverage. Data of this sort are often available from the computerized billing database and therefore, are easy to collect and analyze. As the researcher, you might select patients with a particular diagnosis or length of stay. In this example, we will consider every patient admitted into the hospital, but will exclude patients admitted into the ICU department. A random sample of 100 patients falling into each of the four insurance coverage groups will be selected.

HYPOTHESIS 8.3 – AVERAGE PATIENT CHARGES FOR PAIN RELIEF DIFFER BY INSURANCE COVERAGE

Hypothesis 8.3 has two variables. The first variable is Average Charge for Pain Relief. It is a continuous variable measured in dollars. The second variable is Insurance Coverage. It is a categorical variable measured as one of four groups; private insurance,

Medicaid, Medicare, and no insurance. Table 8-4 displays the computer generated results of the ANOVA.

Table 8-4
Average Charge for Pain Relief by Insurance Coverage
ANOVA and Descriptive Statistics

	Private Insurance	Medicaid	Medicare	No Insurance
Mean (in dollars)	359.10	306.10	309.90	253.90
Standard Deviation	68.06	67.18	70.27	57.01
Sample Size	100	100	100	100

$F = 42.64;\ df = 3;\ p = 0.0000$

The results indicate that patients within the no insurance group are charged the least, $\bar{x} = \$253.90$. Patients in the private insurance group are charged the most, $\bar{x} = \$359.10$. The probability level of 0.0000 is less than the established cut-off of 0.05, thus the results of the ANOVA indicate that there are significant differences in the average patient charges for pain relief by insurance coverage. Therefore, our hypothesis of "Average patient charges for pain relief differ by insurance coverage" is accepted.

To determine where the significant differences exist between these four groups, we will compare the lower and upper 95% confidence intervals printed by the statistics software. The results are displayed in Table 8-5.

Table 8-5
Average Charge for Pain Relief by Insurance Coverage
Lower and Upper 95% Confidence Intervals

	Private Insurance	Medicaid	Medicare	No Insurance
Lower	345.60	292.80	296.00	242.60
Upper	372.60	319.50	323.90	265.20

In studying Table 8-5, we can see that there are many significant differences in pain relief charges between the four groups of patients. Private insurance patients are charged significantly more than the other three groups and no insurance patients are charged significantly less than the other three groups. There are no significant differences in the pain relief charges between the Medicaid and Medicare patients. See Example 8-3 for a sample report of these results.

> **Example 8-3**
> **Sample Report**
>
> The billing records of 400 patients admitted to the hospital over the past year were studied for pain relief charges. The records were randomly selected to include 100 patients each with private insurance, Medicaid, Medicare, and no insurance. Patients admitted into the ICU department were excluded from the study.
>
> The average pain relief charge ranged from \$253.90, for no insurance patients, to \$359.10, for private insurance patients (see Table 8-4). The results of the ANOVA ($F = 42.64$; $p = 0.0000$) indicate that private insurance patients are charged significantly more than the other three groups and no insurance patients are charged significantly less than the other three groups. There are no significant differences in the pain relief charges between the Medicaid and Medicare patients.

A step-by-step guideline for hypothesis testing using analysis of variance can be found in Example 8-4.

> **Example 8-4**
> **Steps in Hypothesis Testing Using ANOVA**
>
> Step 1: Double check that one of your variables divides your data into more than two groups (categorical) and that your second variable is continuous (see Chapter 2).
>
> Step 2: Run the analysis of variance test using your computerized statistical program.
>
> Step 3: Compare the calculated probability (also called significance level) to the standard cut-off of 0.05.
>
> Step 4: If the calculated probability is greater than the 0.05 cut-off, reject your hypothesis and write the appropriate report of no significant differences found.
>
> Step 5: If the calculated probability is less than or equal to the 0.05 cut-off, accept your hypothesis. Follow up with additional testing to determine where the differences were uncovered. Write the appropriate report for your significant differences.

Chapter 9

CORRELATION

A correlation is the appropriate statistic when both of your variables are continuous in nature. Refer to Chapter 2 for information concerning the classification of your variables. While some writers will use the term correlation when they actually mean relationship, the term correlation should only be used to refer to the specific statistical test. The computer generated analysis for the correlation provides a correlation coefficient (r), the sample size (n), and the probability or significance level (p). Most computerized statistical programs provide the correlation results in a table or matrix similar to a mileage or multiplication table.

The correlation tests whether two variables are related to each other. This does not mean that one variable has caused a change in the other, simply that they move in the same or opposite direction. Two variables that move in the same direction will be positively correlated. This means that as variable one increases, variable two increases. A perfect positive correlation coefficient is 1.000. When two variables move in opposite directions, the correlation will be negative. This means that as variable one increases, variable two decreases. A perfect negative correlation coefficient is –1.000. The positive or negative sign will not impact the probability level, but will impact the report you write about your results.

POSITIVE CORRELATION COEFFICIENTS

To some extent, the height of children is related to their age. In general, as children age, they also grow taller. This relationship is not perfect, we have all met some children who are either tall or short for their age. The probability level will report whether the relationship is significant. If the significance level is more than the established cut-off of 0.05, then the two variables are considered unrelated. There is no such thing as 'almost' related. The variables are either significantly related to each other or they are not.

HYPOTHESIS 9.1 – THERE IS A RELATIONSHIP BETWEEN CHILDREN'S AGE AND HEIGHT

Hypothesis 9.1 has two variables, Age and Height. Both Age and Height are continuous variables. Age is measured in years and Height is measured in inches. Since both variables are continuous, the correlation is the appropriate statistic to test this hypothesis. When the software tests this hypothesis, it measures how close to a straight line the data points come. For each child's age, they have a corresponding height. Both the age and height are plotted on a hypothetical graph. The graph is hypothetical because the software does not typically print the graph unless specifically requested to do so. Figure 9-1 displays a scatter plot of the data collected for Hypothesis 9.1. The software then calculates a best fit line for the data set. This line has been drawn in for Figure 9-1. If the data are significantly different from the best fit line, then the probability level will be greater than 0.05.

Most researchers do not prepare the scatter plot displayed in Figure 9-1. It is only shown here to provide a description of what the correlation statistic tests with the data entered. You should prepare a plot graph only if specifically requested to do so.

As we look at Figure 9-1, we can see that there are only a few data points which fall directly or nearly directly on the best fit line. Just because the other data points are off the line does not mean that they are significantly different from the line. The only way to test this significance is to run the correlation statistic. Table 9-1 displays the results of this test.

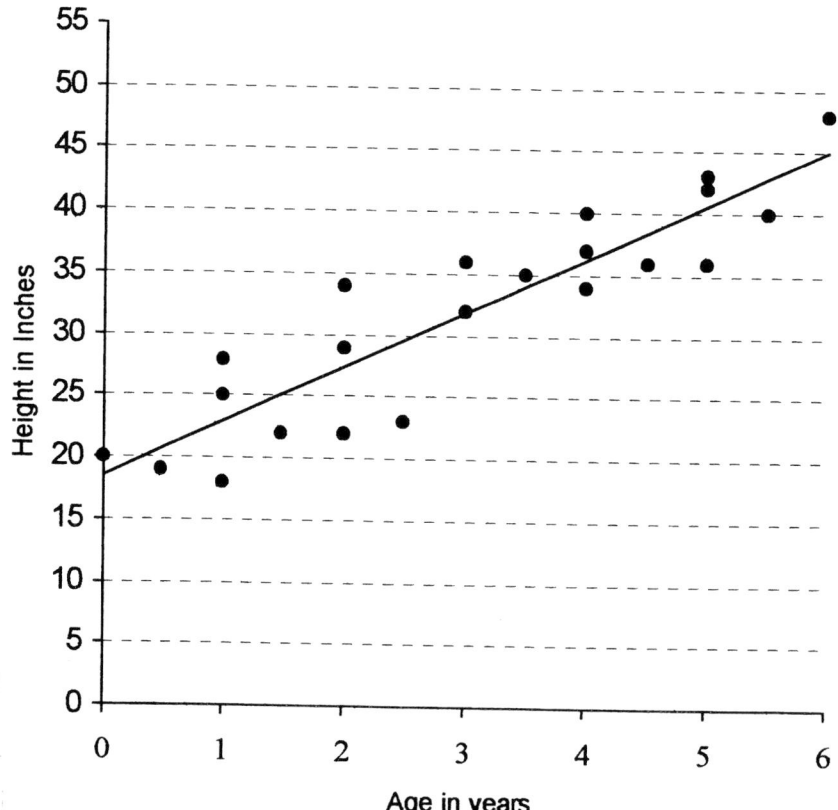

Table 9-1
Children's Age and Height Correlation

	Age	Height
Age	1.000	0.9063
Height	0.9063	1.000

$p = 0.0000; n = 22$

As shown in Table 9-1, the correlation coefficient between age and age is a perfect 1.000. This means that as the child's age increases, the child's age increases. This sounds like nonsense, but we would expect nothing less than a perfect 1.000 correlation coefficient between two identical measures. The same perfect correlation coefficient is also displayed for height and height. The correlation coefficient we are interested in, however, is that of age and height. This correlation coefficient is 0.9063. Notice that it does not matter if you read the table as age and height or height and age. The correlation coefficient is not the deciding factor when it comes to accepting or rejecting Hypothesis 9.1. Remember, we must compare the probability (or significance) level to the established cut-off (see Chapter 5) of 0.05. If the calculated probability is less than or equal to 0.05, then the hypothesis should be accepted. The probability for a correlation coefficient of 0.9063 with a sample size of 22 is 0.0000. Thus, our hypothesis of "There is a relationship between children's age and height" is accepted. The data points which are not directly on the best fit line are not significantly different from the best fit line. The positive sign on the correlation coefficient means that as children's age increases, their height also increases. See Example 9-1 for a sample report based on this data.

Example 9-1
Sample Report

The heights for 22 children between the ages of newborn and 6 years were collected to test the hypothesis that there is a relationship between children's age and height. The results of the correlation indicate that there is a significant positive correlation coefficient (r = 0.9063; p = 0.0000) between the age and height of children in the sample, see Table 9-1.

The acceptance of Hypothesis 9.1 comes as no surprise to us. Based on our experiences, we know that children grow taller as they age. The relationship between other variables, however, is not always so obvious. Let us consider the relationship between length of time waiting for service and diner satisfaction in a restaurant.

HYPOTHESIS 9.2 – THERE IS A RELATIONSHIP BETWEEN SPEED OF SERVICE AND DINER SATISFACTION

Hypothesis 9.2 has two variables. One variable is Speed of Service. It is continuous and is measured in minutes from the time the diner provides the hostess with his/her name, to the time the server presents the bill. The second variable, Diner Satisfaction, is also continuous. It is measured as the number of satisfaction points expressed on an attitude scale. Since both of our variables are continuous, the correlation is the appropriate statistical test of this hypothesis. The results of the correlation are displayed in Table 9-2.

The software program may print the correlation coefficient between the same variables (1.0000) and may print the duplicate correlation coefficients, or it may not. The printout should not be retyped verbatim for your paper. Instead you should only include the relevant information. The descriptive statistics of means and standard deviations are also frequently reported for your readers. Unlike Table 9-1 which displays what might be actually printed by your software, Table 9-2 displays only that information which you might include in your report.

Table 9-2
Diner Satisfaction and Speed of Service
Correlation and Descriptive Statistics

	Mean	Standard Deviation	Correlation Diner Satisfaction
Diner Satisfaction	32.03	11.22	
Speed of Service (in minutes)	46.61	16.14	0.1851*

*$p = 0.3188$; $n = 31$

As shown in Table 9-2, the correlation coefficient between Diner Satisfaction and Speed of Service is 0.1851. Given the sample size of 31, the probability for this correlation is 0.3188. Since this probability is greater than the cut-off of 0.05, we must reject our hypothesis of "There is a relationship between speed of service and diner satisfaction."

Figure 9-2 displays the Scatter Plot of Diner Satisfaction and Speed of Service. We can see by looking at the data that satisfaction is low for long wait times, but also low for short wait times. Thus, the data do not fit the line very well. Example 9-2 provides a sample report based on this data.

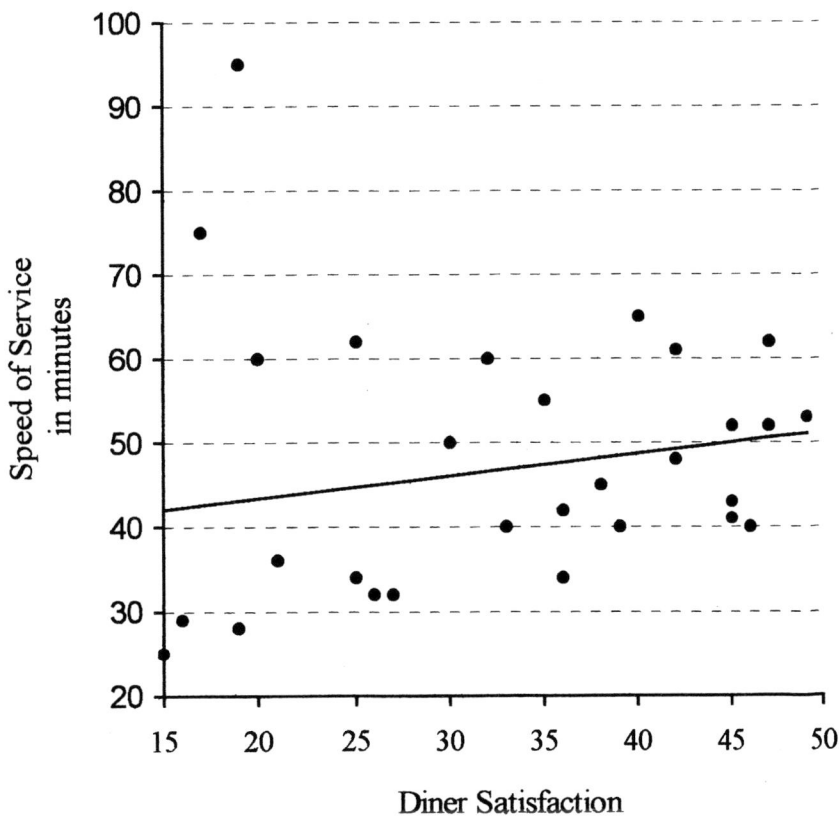

> Example 9-2
> Sample Report
> Speed of service and diner satisfaction measures were gathered for 31 parties who visited The Seafood House during Saturday and Sunday dinner hours for the first weekend in April. Diners were selected at random during check-in. The average speed of service for diners visiting during this weekend was 46.61 minutes. Average diner satisfaction was reported as 32.03 points. The results of the correlation indicate that there is not a significant relationship between speed of service and diner satisfaction ($r = 0.1851$; $p = 0.3188$), see Table 9-2.

NEGATIVE CORRELATION COEFFICIENTS

Employee Assistance Programs frequently offer programs to reduce the incidence of stress symptoms among employees. Unfortunately, few companies actually test the effectiveness of such programs. If these programs are successful in significantly reducing the symptoms of stress among employees, then their costs can be justified. However, if the programs are not successful, then changes should be made.

HYPOTHESIS 9.3 – THE USE OF EMPLOYEE ASSISTANCE EXERCISE PROGRAMS IS RELATED TO LOWER REPORTED STRESS SYMPTOMS

Hypothesis 9.3 has two variables. One variable is the use of the Exercise program. It is a continuous variable measured in hours of use per week. The second variable is Stress Symptoms. It is a continuous variable measured as cumulative points on a self-report attitude scale. Since both of our variables are continuous, the correlation is the appropriate test of this hypothesis. Table 9-3 displays the results of the correlation of Exercise and Stress Symptoms.

Table 9-3

Exercise and Stress Symptoms
Correlation and Descriptive Statistics

	Mean	Standard Deviation	Correlation Exercise
Exercise (in hours)	0.7538	0.5695	
Stress Symptoms	68.15	19.21	-0.5097*

*$p = 0.0752$; $n = 13$

As shown in Table 9-3, the results indicate that there is a negative correlation coefficient of -0.5097 for Exercise and Stress Symptoms. Figure 9-3 displays the scatter plot for the data so that you may visualize the negative relationship. The negative sign means that as the number of hours of Exercise per week increases, the reported Stress

Symptoms decreases. This sound good, except that the probability for this correlation coefficient with a sample size of 13 is 0.0752. This probability is greater than the cut-off of 0.05, so we must reject our hypothesis. There is not a significant relationship between the use of Employee Assistance Exercise programs and a reduction in reported Stress Symptoms. The probability is close to the cut-off, but close is not good enough. A sample report for Hypothesis 9.3 appears in Example 9-3.

The sample size in this example is quite small (only 13). To determine the probability level, the computer uses the strength of the correlation coefficient and the sample size. Thus, with a larger sample size, this correlation coefficient would be sufficiently strong to accept the hypothesis. If possible, larger sample sizes should be utilized. It would be permissible for the researcher in this situation to continue to gather data to increase the sample size. As long as the researcher does not specifically select or reject cases to include or exclude (this would be unethical), re-testing the hypothesis with a larger sample size would be acceptable. For a step-by-step guideline for hypothesis testing using the correlation, see Example 9-4.

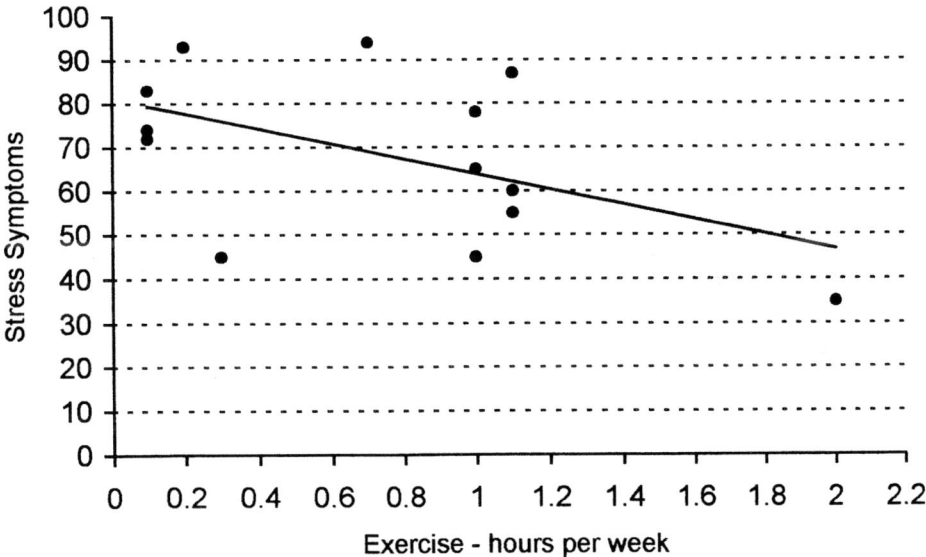

> Example 9-3
> Sample Report
>
> Thirteen employees participating in the Employee Assistance exercise program were surveyed to determine their current level of stress symptoms. Average reported stress symptoms were 68.15 (SD = 19.21). Sign-in sheets in the exercise room were reviewed for the previous three week period to determine the average number of hours each employee exercised per week. On average, employees exercised 0.7538 hours per week (about 45 minutes). The results of the correlation indicate that there is not a significant relationship between the use of Employee Assistance exercise programs and reported stress symptoms ($r = -0.5097$; $p = 0.0752$), see Table 9-3.

> Example 9-4
> Steps In Hypothesis Testing Using Correlation
>
> Step 1: Double check that both of your variables are continuous (see Chapter 2).
>
> Step 2: Run the correlation test using your computerized statistical program.
>
> Step 3: Note the sign of the correlation coefficient.
>
> Step 4: Compare the calculated probability (also called significance level) to the standard cut-off of 0.05.
>
> Step 5: If the calculated probability is greater than the 0.05 cut-off, reject your hypothesis and write the appropriate report of no significant relationship found.
>
> Step 6: If the calculated probability is less than or equal to the 0.05 cut-off, accept your Hypothesis. Write the appropriate report for your significant relationship. Note the direction of the relationship in your report. Use caution not to mention a causal relationship between the two variables.

Chapter 10

TESTING YOUR SCALES

If you have used attitude measures to collect your data, you must demonstrate that these measures have reliability, validity, and sensitivity. If you have collected data as a single measure of some concrete value (sales, weight, attendance, etc.), then reliability and validity need not be addressed in your report.

RELIABILITY

Test-retest reliability – Test-retest reliability means that you can depend upon your measure to give you similar answers over repeated uses. If I were to measure your job satisfaction today and in two weeks, and there had not been any change in your actual job satisfaction, then the score should not be significantly different for the first and second measurement.

Test-retest reliability is important during scale development. If you have designed your own scale, you should be able to demonstrate test-retest reliability. In order to demonstrate test-retest reliability, you must gather data from the same sample during two periods in time. The subjects should place some type of identification code on both surveys so that the first survey can later be matched with the second survey. The sample may or may not be the same group for your hypothesis testing. Strive to have a sample size of at least 50 subjects in your test-retest group. A two week interval between test days is typically sufficient. The data from the first session should be compared with the data from the second session using a correlation. You should expect that the correlation coefficient for the two measures of job satisfaction would be extremely high. Correlation coefficients should be at least 0.80 for your measures to be considered reliable over time. Probability values are not considered when demonstrating test-retest reliability. See Example 10-1 for a sample report of a test-retest reliability measure.

> Example 10-1
> Sample Report
>
> Three classroom groups of students enrolled in a Master's Program in Human Resources were given a job satisfaction survey to complete in class. Participation was voluntary. All students were employed full-time and were attending classes in a weekend program. Two weeks later, students were again asked to complete the same survey. A total of 53 students completed both surveys. Twelve students were only present for one survey date and their surveys were excluded from the analysis.
>
> In order to determine the test-retest reliability of the job satisfaction survey, a correlation coefficient was calculated. The results indicate that the job satisfaction scale demonstrated sufficient test-retest reliability ($r = 0.8704$) for use in hypothesis testing.

If your scale is not reliable over time, then you should not use it. Imaging using a bathroom scale that gave a different weight each time you stepped on it. You could not depend on this scale and you would not use it. This makes sense. If the scale cannot be depended upon to provide the same answer with each use (assuming no changes to the underlining value), then it is worthless as a research tool. A new scale should be selected. If you have not developed the survey yourself, but rather are using a previously developed scale with a positive history of psychometric testing, then it is not necessary for you to demonstrate test-retest reliability. This is one of the many benefits of using previously validated measurement tools.

Internal reliability – Internal reliability or internal consistency is important to demonstrate when your scales are a summation of more than one item. Frequently we measure attitudes on a multiple item survey. If I were to ask you about your job satisfaction, my survey might include ten items about job satisfaction. Your satisfaction would be an average or a summation of these items. It is important for the researcher to demonstrate that these ten items all measure the same thing, job satisfaction. If a few of the items were found to be a measure of job commitment instead, for example, then these items should not be included in the measure of job satisfaction.

Most statistical software programs will calculate the internal reliability of your scale using the Cronbach's coefficient alpha. The coefficient alpha should be at least 0.70 for your scales to be considered sufficiently internally reliable for use in hypothesis testing. The coefficient alpha (printed as α or alpha) should be calculated and reported for each multiple-item scale you have used, whether you have developed the scale yourself or you have used a previously existing one.

When using previously developed scales, in most cases the alpha level will meet the minimum level of acceptability. This is another benefit of previously validated measurement tools. If it does not, care should be taken to make sure you have properly reversed scored (see Chapter 5) any items coded as such by the test designer. Besides ensuring that you have properly coded the surveys, there is little else you can do to improve the internal consistency of scales designed by someone else. Checking for high reports of internal reliability before you select a scale may help ensure a good test for your hypotheses. Table 10-1 displays information which may be printed by your software when calculating Cronbach's alpha for internal reliability.

Table 10-1
Job Satisfaction
Internal Reliability

Item	Scale Mean if Item Deleted	Scale Variance if Item Deleted	Corrected Item Total Correlation	Alpha if Item Deleted
1	43.3446	4.9093	0.5641	0.6083
2	43.1311	4.2307	0.4916	0.6398
3	44.1341	4.5432	0.5395	0.6665
4	43.0141	3.7254	0.2157	0.7654
5	42.4320	4.3325	0.4666	0.6103
6	42.3234	4.5332	0.3509	0.6551
7	43.1311	3.7354	0.5064	0.6402
8	43.0340	3.5744	0.2967	0.6589
9	44.2322	3.7454	0.5454	0.5474
10	42.2342	3.8585	0.4474	0.6951
11	43.9242	3.9259	0.5916	0.6102
12	41.4329	3.8329	0.4365	0.6859
13	42.3234	4.9334	0.2667	0.7985
14	44.2322	3.8353	0.3666	0.7751
15	43.9349	3.8745	0.5009	0.5982

$n = 48$; alpha = 0.6875

If the scales you have designed produce an alpha lower than 0.70, you should study the computerized results for items which may not be a good fit. As shown in Table 10-1, the alpha reported of 0.6875 is less than 0.70. Oftentimes the software will suggest which item should be removed from the scale because it is not a good fit with the other items.

The last column of Table 10-1 displays the alpha if the item is deleted. Items 4, 13, and 14 are indicated to be items which may not be a good fit. This is shown by the increase in alpha above 0.70 if the item were to be removed. One item should be removed at a time and the alpha recalculated. When the alpha is reported as 0.70 or above, the remaining scale items should be retained. If removing a few items from your scale improves your alpha sufficiently, the new, shorter scale can then be used for hypothesis testing. You should discuss the change in scale length in your reporting. See Example 10-2 for a sample report.

> **Example 10-2**
> **Sample Report**
>
> A 15-item scale was designed to measure the subjects' level of job satisfaction. Initial testing for internal reliability indicated that three items were not a good fit with the others. These three items were removed from further analysis (see the appendix for a copy of the final survey). The 12-item satisfaction survey produced a Cronbach's alpha of 0.79. The average satisfaction level for the sample of 48 employees was 42.45 (SD = 2.12).

If the internal reliability measures do not meet the established 0.70 cut-off and corrections to coding or scale length are ineffective, then ideally the scale should not be used in hypothesis testing. The researcher should select or design a new scale and repeat the data gathering process. Poorly designed scales may lead to the rejection of hypotheses which may actually be true. If the scales must be utilized (perhaps because of academic expectations), the paper should include an explanation of the limitations of the scales. Example 10-3 provides a sample report describing such scale limitations.

> **Example 10-3**
> **Sample Report**
>
> The results of this study are limited by the poor internal reliability of the job satisfaction scale ($\alpha = 0.5423$). It is recommended that future research in this area utilize a job satisfaction measure with stronger psychometric properties.

VALIDITY

Validity means that the scale is actually measuring what you intended to measure. We know that when we step on a bathroom scale, we are measuring our weight and not our height. Sometimes validity is as easy as looking to see if the scale is actually measuring what we intend. This is termed face validity or content validity. Face validity is easier to demonstrate for a bathroom scale than it is for most attitude measures. If you are measuring job satisfaction, your items should actually include the word satisfaction and should all be focused on the aspects of someone's job. Asking someone if they like their company or even if they like their job is not a measure of job satisfaction. While liking may mean the same thing as satisfaction to you, it may not mean the same thing to each person in your study. Likewise, employees may be satisfied with their job, but be dissatisfied with the company they work for.

For scale development, researchers typically demonstrate that their attitude measure is valid by comparing the results obtained from their measure to that of a previously designed scale considered to be valid. If you have selected a scale that has been previously tested, then demonstrating validity is not typically necessary. As we see, the benefits of using previously designed and tested scales are numerous.

SENSITIVITY

Scales should be sensitive enough to pick up on changes if changes have occurred. If your job satisfaction was measured on a two point scale, *yes* or *no*, then it is unlikely we would be able to measure slight changes in your attitude about your job satisfaction. If you are satisfied today and tomorrow you receive a 20% pay raise, you are satisfied tomorrow too. Most of us would be more satisfied with a 20% pay raise, but the yes/no scale does not allow for the increase in satisfaction to be measured. Frequently attitude measures include a 5 or 7 point scale. If you use many more than 7 values in your scale, it may be difficult for the subjects to distinguish between the values. Fewer than 5 values may not be sensitive enough to pick up on slight changes in the subject's attitude.

Your written report should describe each of your measurements. You should include the psychometric properties of the measurements (reliability and validity) and the unit of measurement. See Example 10-4 for a sample measurement report.

Example 10-4
Sample Report

Organizational Satisfaction is the extent to which respondents are satisfied with their current organization. Three organizational satisfaction items were developed for this study. The items were measured on a five-point scale ranging from Very dissatisfied (1) to Very satisfied (5). For this sample (n = 68), the Organizational Satisfaction scale produced a Cronbach's alpha of 0.85. A prior independent test sample of 57 respondents completed the organizational satisfaction measure at two different times, two weeks apart. The test-retest reliability correlation coefficient of 0.8125 indicated sufficient reliability for the organizational satisfaction measure to be used in hypothesis testing.

LIST OF HYPOTHESES

Hypothesis 2.1 – Employee location is related to level of satisfaction.   7

Hypothesis 2.2 – Learning time is related to age and gender.   9

Hypothesis 2.2a – Learning time is related to age.   9

Hypothesis 2.2b – Learning time is related to gender.   9

Hypothesis 6.1 – Coin Toss is unfair.   29
$\chi^2 = 0.0909$; $df = 1$; $p = 0.7630$; Rejected

Hypothesis 6.2 – Coin toss by coin type is unfair.   30
$\chi^2 = 4.96$; $df = 2$; $p = .0839$; Rejected

Hypotheses 6.3 – Re-employment status is related to the type of therapy received.   31
$\chi^2 = 17.48$; $df = 4$; $p = 0.0016$; Accepted

Hypothesis 6.4 – Testing referrals differ by elementary school.   33
$\chi^2 = 6.274$; $df = 8$; $p = 0.6166$; Rejected

Hypothesis 7.1 – Midterm and final exam scores for students in section 102 are significantly different from each other.   37
$t = -1.72$; $df = 32$; $p = 0.0952$; Rejected

Hypothesis 7.2 – Average employee sick leave used in year 1 is different from the average employee sick leave used in year 2.   39
$t = 4.59$; $df = 209$; $p = 0.0000$; Accepted

Hypothesis 7.3 – Hotel employee organizational commitment will increase following supervisory leadership training.   40
$t = -13.92$; $df = 104$; $p = 0.0000$

Hypothesis 7.4 – Children who participate in daily recess play will express greater school satisfaction than children who do not participate in recess play.   42
$t = 1.95$; $df = 71$; $p = 0.0274$; Accepted

Hypothesis 8.1 – Interest charges for new car loans differ by lending institution. 45
    $F = 5.26$; $df = 3$; $p = 0.0022$; Accepted

Hypothesis 8.2 – There is a relationship between payment method and purchases made. 47
    $F = 0.88$; $df = 2$; $p = 0.4166$; Rejected

Hypothesis 8.3 – Average patient charges for pain relief differ by insurance coverage. 48
    $F = 42.64$; $df = 3$; $p = 0.0000$

Hypothesis 9.1 – There is a relationship between children's age and height. 52
    $r = 0.9063$; n = 22; $p = 0.0000$; Accepted

Hypothesis 9.2 – There is a relationship between speed of service and diner satisfaction. 54
    $p = 0.3188$; $n = 31$; Rejected

Hypothesis 9.3 – The use of Employee Assistance exercise programs is related to lower reported stress levels. 56
    $r = -0.5097$; $n = 13$; $p = 0.0752$; Rejected

EXERCISES

CHAPTER 2 – CLASSIFYING YOUR DATA

1. Classify the following variables as categorical with a two-value limit, categorical with no value limit, or continuous.
    a. classroom
    b. income
    c. teacher
    d. bank branch
    e. travel distance
    f. weight
    g. nationality
    h. agreement (yes/no)
    i. agreement (strongly agree to strongly disagree)
    j. batting average
    k. toothpaste brand

2. Write two bivariate hypotheses using the variables listed in question 1. State the appropriate statistical technique (using Table 2-1) for testing each.

3. Rewrite the following multivariate hypotheses into bivariate hypotheses.

    a. Brand Loyalty is related to length of time using the product and coupon availability.
    b. Batting average and income are related to experience in college and in the minors.
    c. Employee satisfaction is related to gender, shift, and supervisor experience.

4. After rewriting the hypotheses in question 3, state the statistical technique appropriate for testing each of the bivariate hypotheses.

CHAPTER 3 – FREQUENCY DISTRIBUTIONS

1. Write a report based on the data recorded in Table 3-1.

2. Prepare a frequency distribution for the following data on music preferences. Write a report based on this distribution. The data has been coded as 1 – rock and roll, 2 – pop, 3 – classical, 4 – rap, 5 – country and western, 6 – rhythm and blues, and 7 – other.

   1 4 2 5 2 4 2 5 1 7 2 7 2 1 5 2 4 3 6 2 1 4 2 5 6 3 7 2 1 6 4 2 5 2 4 2 5 2 1 1 1 2
   2 1 5 2 4 3 6 2 1 4 2 5 6 3 7 2 5 1 7 2 7 2 1 5 2 2 5 2 4 2 5 5 2 1 1 1 2 3 6 6 3 1

3. Write a report based on the data recorded in Table 3-4.

4. Prepare a cross-tabulation table for the data on music preferences listed in question 2. Assume the first 42 observations are from male respondents and the second 42 observations are from female respondents. Write a report based on this cross-tabulation.

CHAPTER 4 – DESCRIPTIVE STATISTICS

1. Study the data displayed in Table 4A. Determine the appropriate descriptive statistics from the table. Write a report based on this information.

Table 4A
Long Distance Carrier
Frequency Distribution

Long Distance Carrier	Frequency	Cumulative Frequency	Percentage	Cumulative Percentage
AT&T	18	18	23%	23%
MCI	25	43	32%	55%
Sprint	17	60	22%	77%
Other	18	78	23%	100%
Total	78	78	100%	100%

2. Study the data displayed in Table 4B. Determine the appropriate descriptive statistics from the table. Write a report based on this information.

Table 4B
Weight of Potato Chips in a 12 Ounce Bag
Frequency Distribution and Descriptive Statistics

Weight in ounces	Frequency	Cumulative Frequency	Percentage*	Cumulative Percentage
10.50	3	3	07	07
10.75	1	4	02	09
11.00	4	8	09	18
11.25	1	9	02	20
11.50	5	14	11	31
11.75	10	24	22	53
12.00	11	35	24	78
12.25	4	39	09	87
12.50	3	42	07	93
12.75	2	44	04	98
13.00	1	45	02	100
Total	45	45	100%	100%

*May not sum to 100%, due to rounding. $\bar{x} = 11.77$; $SD = 0.5921$.

3. Write a report based on the data recorded in Table 4-2, include the information $\bar{x} = 6.12$; $SD = 2.45$.

4. Write a report based on the data recorded in Table 4-3, include the information $\bar{x} = \$1018$; $SD = \$86.23$.

CHAPTER 5 – HYPOTHESIS TESTING

1. Determine if you should accept or reject your hypothesis for each probability level printed.

   a. 0.4589
   b. 0.8521
   c. 0.0924
   d. 0.0097
   e. 0.4001
   f. 0.0638
   g. 0.9999
   h. 0.0099
   i. 0.0501
   j. 0.1002

2. Prepare coding values for each of the following categorical variables.

   a. Gender – male, female
   b. Shift – first, second, third
   c. Department – accounting, marketing, human resources, research and development, manufacturing
   d. Working – full-time, part-time, not at all
   e. Education – some high school, high school graduate, some college, college graduate, post graduate

3. The following five items are part of an attitude scale. Study the scale to determine if any of the items should be reverse coded. Write the codes that should be entered into the computer for this subject. What is this subject's total score for this survey? Write a report about this scale and the recoding you recommend.

	1 Strongly disagree 2 Disagree 3 Neither disagree or agree 4 Agree 5 Strongly agree				
	1	2	3	4	5
My supervisor cares about the employees.				X	
My supervisor considers employees important.					X
My supervisor ignores employees when they have something to say.		X			
I consider my supervisor easy to talk to.			X		
I do not feel comfortable disagreeing with my supervisor.	X				

4. Determine if each of the following hypotheses should be tested with a one-tail or a two-tailed test.

   a. Employees working at Store 1 earn higher salaries than employees at the other stores.
   b. IBM compatible computers are more popular than Macintosh computers.
   c. Customers prefer grocery stores with speed lane service over stores without such service.
   d. There is a relationship between employee satisfaction and absences.
   e. Female customers pay higher dry cleaning rates.
   f. Car prices differ between the dealerships in the two cities.
   g. Contests encourage magazine subscription renewal decisions.
   h. Music preferences differ by age.

CHAPTER 6 – CHI-SQUARE

1. Consider the following hypothesis and table of results. Determine if the hypothesis should be accepted or rejected. Write a report based on these results.

Hypothesis 6A – Plans to pursue graduate accounting education differ based on student program of enrollment.

Table 6A
Graduate Education Plans by Program
Cross Tabulation and Chi-square

Program	Plans			Total
	Within 1 year	Within 5 years	No plans	
Day	12	2	2	16
Career	20	5	1	26
Total	32	7	3	42

$\chi^2 = 1.3125$, $df = 2$, $p = 0.5188$

2. Consider the following hypothesis and table of results. Determine if the hypothesis should be accepted or rejected. Write a report based on these results.

Hypothesis 6B – Marital status is related to decision to purchase a time-share vacation plan.

Table 6B
Purchase Decision by Marital Status
Cross Tabulation and Chi-square

Marital Status	Purchase		Total
	Yes	No	
Married	14	31	45
Single	8	13	21
Total	22	44	66

$\chi^2 = 0.31$, $df = 1$, $p = 0.5751$

3. Consider the following hypothesis and table of results. Determine if the hypothesis should be accepted or rejected. Write a report based on these results.

Hypothesis 6C – There is a relationship between day of the week employees call the human resource help desk and topic of help requested.

Table 6C
Type of Call by Day of the Week
Cross Tabulation and Chi-square

Type of Call	Day of the Week					Total
	Mon	Tues	Wed	Thurs	Fri	
Insurance	3	5	7	2	0	17
Payroll	15	20	2	10	4	51
Sick Leave	27	26	7	5	4	69
Job Opportunities	13	6	2	6	5	32
Total	58	57	18	23	13	169

$\chi^2 = 31.5220, df = 12, p = 0.0016$

4. Consider the following hypothesis and table of results. Determine if the hypothesis should be accepted or rejected. Write a report based on these results.

Hypothesis 6D – There is a relationship between employee gender and voluntary attendance at a Sexual Harassment Awareness Lecture.

Table 6D
Lecture Attendance by Employee Gender
Cross Tabulation and Chi-square

Gender	Attendance		Total
	Yes	No	
Male	25	0	25
Female	9	5	14
Total	34	5	39

$\chi^2 = 10.24, df = 1, p = 0.00137$

CHAPTER 7 – *t* TESTS OF TWO MEANS

1. Determine if the following hypotheses should be tested using a paired or independent *t* test and whether a one- or two-tailed test should be used.

    a. Employee satisfaction will increase following a pay increase.
    b. The tendency for employees to report wrongdoing is related to years of experience on the job.
    c. The completion of manager evaluations will be higher for employees with multiple year contracts.
    d. Customer incentives for free air-time are related to cellular phone service contract renewals.

2. Consider the following hypothesis and table of results. Determine if the hypothesis should be accepted or rejected. Write a report based on these results.

    Hypothesis 7A – Faculty and student views differ concerning academic dishonesty.

    Table 7A
    Faculty and Student Views Concerning Academic Dishonesty
    t Test and Descriptive Statistics

	Faculty	Student
Mean	3.4390	2.6970
Standard Deviation	1.184	1.359
Sample Size	41	132

    $t = 3.14$; $df = 171$; $p = 0.002$

3. Consider the following hypothesis and table of results. Determine if the hypothesis should be accepted or rejected. Write a report based on these results.

    Hypothesis 7B – Employee contract term is related to union membership status.

Table 7B
Employee Contract Term by Union Membership Status
t Test and Descriptive Statistics

	Union Member	Not a Union Member
Mean (in years)	3.8222	4.1860
Standard Deviation	1.232	1.101
Sample Size	90	86

$t = -2.06; df = 174; p = 0.041$

4. Consider the following hypothesis and table of results. Determine if the hypothesis should be accepted or rejected. Write a report based on these results.

Hypothesis 7C – The use of a quality-control technician will lead to a reduction in the number of parts returned due to manufacturing defects.

Table 7C
Parts Returned With and Without Quality Control
One-tailed t Test and Descriptive Statistics

	Without Quality Control	With Quality Control
Mean (per thousand)	2.0400	1.6818
Standard Deviation	1.306	1.235
Sample Size	25	44

$t = 1.13; df = 67; p = 0.131$

CHAPTER 8 – ANOVA ANALYSIS OF VARIANCE

1. Consider the following hypothesis and tables of results. Determine if the hypothesis should be accepted or rejected. Write a report based on these results.

Hypothesis 8A – Combined waiting and treatment time for patients is related to the day of the week of the appointment.

Table 8A1
Wait/Treatment Time by Day of Week
ANOVA and Descriptive Statistics

	Day of the Week				
	Mon	Tues	Wed	Thurs	Fri
Mean*	59.18	52.18	56.89	61.30	69.91
SD	13.24	20.43	18.09	10.55	14.72
n	11	11	19	10	32

*in minutes; $F = 3.6391$; $df = 4$; $p = 0.0090$

Table 8A2
Wait/Treatment Time by Day of Week
Lower and Upper 95% Confidence Intervals

	Day of the Week				
	Mon	Tues	Wed	Thurs	Fri
Lower	51.36	40.11	48.76	54.76	64.81
Upper	67.01	64.26	65.03	67.84	75.01

2. Consider the following hypothesis and tables of results. Determine if the hypothesis should be accepted or rejected. Write a report based on these results.

Hypothesis 8B – Student class is related to time of persistence with difficult tasks.

Table 8B1
Time of Persistence by Student Class
ANOVA and Descriptive Statistics

	Student Class				
	Freshman	Sophomore	Junior	Senior	Graduate
Mean*	2.56	2.61	2.98	2.73	3.44
SD	1.12	1.44	1.34	1.40	1.18
n	25	23	48	26	41

*in minutes; $F = 2.6087$; $df = 4$; $p = 0.0377$

Table 8B2
Time of Persistence by Student Class
Lower and Upper 95% Confidence Intervals

	Student Class				
	Freshman	Sophomore	Junior	Senior	Graduate
Lower	2.10	1.99	2.59	2.17	3.07
Upper	3.02	3.23	3.37	3.30	3.81

3. Consider the following hypothesis and tables of results. Determine if the hypothesis should be accepted or rejected. Write a report based on these results.

Hypothesis 8C – Employee satisfaction is related to shift.

Table 8C1
Employee Satisfaction by Shift
ANOVA and Descriptive Statistics

	Shift		
	First	Second	Third
Mean	67.80	66.81	71.00
SD	13.80	13.67	12.00
n	69	36	30

$F = 0.8815; df = 2; p = 0.4166$

Table 8C2
Employee Satisfaction by Shift
Lower and Upper 95% Confidence Intervals

	Shift		
	First	Second	Third
Lower	64.48	62.18	66.52
Upper	71.11	71.43	75.48

4. Consider the following hypothesis and tables of results. Determine if the hypothesis should be accepted or rejected. Write a report based on these results.

Hypothesis 8D – Book prices differ by seller.

Table 8D1
Book Prices by Seller
ANOVA and Descriptive Statistics

	Seller			
	Chain	Independent	Online	University
Mean*	40.40	39.90	41.40	39.40
SD	15.56	16.80	16.38	15.90
n	10	10	10	10

*in dollars; $F = 0.03$; $df = 3$; $p = 0.9936$

Table 8D2
Book Prices by Seller
Lower and Upper 95% Confidence Intervals

	Seller			
	Chain	Independent	Online	University
Lower	29.97	27.88	29.68	28.02
Upper	51.53	51.92	53.12	50.78

CHAPTER 9 – CORRELATION

1. Study the correlation matrix displayed in Table 9A. Write three hypotheses based on the variables listed and state whether they are accepted or rejected. Write the appropriate reports for your hypotheses and prepare appropriate tables for your readers.

Table 9A
Employee Attitudes
Intercorrelations (Probabilities), Descriptive Statistics

	Mean	SD	1	2	3	4
1. Satisfaction	69.609	12.596	1.00	-0.0625 (0.1810)	-0.5988 (0.0000)	-0.5481 (0.0000)
2. Continuance Commitment	34.363	8.908	-0.0625 (0.1810)	1.00	-0.0199 (0.3840)	-0.1772 (0.0050)
3. Role Conflict	13.252	5.413	-0.5988 (0.0000)	-0.0199 (0.3840)	1.00	0.4940 (0.000)
4. Intention to Quit	11.310	5.761	-0.5481 (0.0000)	-0.1772 (0.0050)	0.4940 (0.000)	1.00

$n = 225$; because of missing data, some statistics were computed with a somewhat smaller sample.

2. Study the correlation matrix displayed in Table 9B and determine if the following hypotheses should be accepted or rejected. Write the appropriate reports for each hypothesis based on these results.

   a. There is a relationship between patient age and intention to remain clean.
   b. There is a relationship between patient age and therapist success rating.
   c. There is a relationship between patient age and days in treatment.
   d. There is a relationship between days in treatment and therapist success rating.
   e. There is a relationship between intention to remain clean and therapist success rating.
   f. There is a relationship between days in treatment and intention to remain clean.

Table 9B
Substance Abuse Admissions
Intercorrelations (Probabilities), Descriptive Statistics

	Mean	SD	1	2	3	4
1. Age	32.525	7.637	1.00	0.2091 (0.053)	-0.2096 (0.052)	-0.3286 (0.005)
2. Days in Treatment	30.803	10.463	0.2091 (0.053)	1.00	0.5067 (0.000)	0.5227 (0.000)
3. Therapist Success Rating	75.333	12.494	-0.2096 (0.052)	0.5067 (0.000)	1.00	0.6615 (0.000)
4. Intention to Remain Clean	8.545	5.106	-0.3286 (0.005)	0.5227 (0.000)	0.6615 (0.000)	1.00

$n = 66$; because of missing data, some statistics were computed with a somewhat smaller sample.

CHAPTER 10 – TESTING YOUR SCALES

1. If you are developing your own scales, which psychometric tests should be completed prior to hypothesis testing?

2. What are the benefits of using previously developed and validated scales?

3. Study the results of the test-retest reliability statistics presented below. Make appropriate recommendations.

Table 10A
Intrapersonal Role Conflict
Test-retest Reliability

	Mean	SD	Correlation Test
Test	19.36	5.81	
Retest	19.03	5.92	0.8085

$n = 77$

4. Study the results of the Cronbach's internal reliability statistics presented below. Make appropriate recommendations.

Table 10B
Continuance Commitment
Internal Reliability

Item	Scale Mean if Item Deleted	Scale Variance if Item Deleted	Corrected Item Total Correlation	Alpha if Item Deleted
1	28.5246	71.8536	0.4651	0.7302
2	28.7377	67.4967	0.5918	0.7063
3	29.7541	74.7552	0.4394	0.7354
4	28.0147	61.1254	0.2147	0.8425
5	29.4590	67.6525	0.5688	0.7103
6	29.3934	72.9093	0.3409	0.7551
7	28.7377	74.2301	0.4065	0.7402
8	28.0820	76.5432	0.2987	0.7589

$n = 61$; alpha = 0.6875

SELECTED ANSWERS TO EXERCISES

CHAPTER 2 – CLASSIFYING YOUR DATA

1.
    a. categorical with no value limit
    b. continuous
    c. categorical with no value limit
    d. categorical with no value limit
    e. continuous
    f. continuous
    g. categorical with no value limit
    h. categorical with a two-value limit
    i. continuous
    j. continuous
    k. categorical with no value limit

2. Student hypotheses will vary.
    Average customer income differs by bank branch. ANOVA
    Nationality of customers are different by bank branch. chi-square

3.& 4.   a. Brand Loyalty is related to length of time using the product. correlation
    Brand Loyalty is related to coupon availability. *t* test

    b. Batting average is related to experience in college. correlation
    Batting average is related to experience in the minors. correlation
    Income is related to experience in college. correlation
    Income is related to experience in the minors. correlation

    c. Employee satisfaction is related to gender. *t* test
    Employee satisfaction is related to shift. ANOVA
    Employee satisfaction is related to supervisor experience. correlation

CHAPTER 3 – FREQUENCY DISTRIBUTIONS

1. Sample report follows. Student reports will vary.

   A total of 50 respondents completed the survey. They ranged in age from 18 to 66 years old, however only 10% of the sample reported their age as over 40 years (see Table 3-1).

2. Sample report follows. Student reports will vary. Tables should be relatively standard.

   Eighty-four people responded to the interviewer concerning their preferences for music. Of these, 27 (32%) indicated that they preferred pop music and 16 (19%) indicated rock and roll as their most preferred. Country and western music was indicated as preferred by 13 (15%) respondents, see Table 3A.

   Table 3A
   Music Preferences
   Frequency Distribution

Type of Music Preferred	Frequency	Percentage*
1. rock and roll	16	19%
2. pop	27	32%
3. classical	6	7%
4. rap	9	11%
5. country and western	13	15%
6. rhythm and blues	7	8%
7. other	6	7%
Total	84	100%

   *May not sum to 100% due to rounding

4. Sample report follows. Student reports will vary. Tables should be relatively standard.

   Forty-two male and 42 female customers responded to the interviewer's request for their music preferences. Of these, 27 (14 males and 13 females) customers indicated that they preferred pop music and 16 (8 males and 8 females) customers indicated rock and roll as their most preferred. Country and western music was indicated as preferred by 13 (6 males and 7 females) customers, see Table 3B.

Table 3B
Music Preferences by Gender
Cross Tabulation

Type of Music Preferred	Gender		Total
	Male	Female	
1. rock and roll	8	8	16
2. pop	14	13	27
3. classical	2	4	6
4. rap	6	3	9
5. country and western	6	7	13
6. rhythm and blues	3	4	7
7. other	3	3	6
Total	42	42	84

CHAPTER 4 – DESCRIPTIVE STATISTICS

1. Mode = MCI. Since Long Distance Carrier is a categorical variable, the mode is the only descriptive statistic appropriate to report. Cumulative frequencies and cumulative percentages may be printed by the software, but should not be included in the tables you prepare for your reader when the variables are categorical.

   Sample report follows. Student reports will vary.

   A total of 78 telephone customers were contacted and asked to report their long distance carrier. MCI was reported as the most frequent carrier (32%). AT&T and Sprint were reported by 23% and 22% of the sample. See Table 4A.

2. Mean = 11.77, Median = 11.75, Mode = 12.00, Range = 10.50 to 13.00, Variance = 0.3506, Standard Deviation = 0.5921.

   Sample report follows. Student reports will vary.

   Forty-five, 12 ounce bags of potato chips were weighed to determine if their actual weight matched that of the advertised weight. The bags weighed between 10.50 ounces and 13.00 ounces, with an average weight of 11.77 ounces. Thirty-five percent of the bags weighed exactly 12 ounces. Over 70% of the bags weighed within ½ of an ounce of the weight advertised.

CHAPTER 5 – HYPOTHESIS TESTING

1.
    a. 0.4589 – Reject
    b. 0.8521 – Reject
    c. 0.0924 – Reject
    d. 0.0097 – Accept
    e. 0.4001 – Reject

2. Student coding values may vary.

    a. 1 = male, 2 = female
    b. 1 = first, 2 = second, 3 = third
    c. 1 = accounting, 2 = marketing, 3 = human resources, 4 = research and development, 5 = manufacturing
    d. 1 = full-time, 2 = part-time, 3 = not at all
    e. 1 = some high school, 2 = high school graduate, 3 = some college, 4 = college graduate, 5 = post graduate

3. Items 3 and 5 should be recoded. For item 3, 2 = 4. For item 5, 1 = 5. Codes for this subject are 4, 5, 4, 3, 5. Total score = 21. Sample report follows. Student reports will vary.

    A five item questionnaire was designed to measure employees' attitudes about their supervisor. The items were coded from 1 *strongly disagree* to *5 strongly agree*. Higher scores are an indication of positive attitudes. Two items were reverse scored.

4.
    a. one-tail
    b. one-tail
    c. one-tail
    d. two-tail

CHAPTER 6 – CHI-SQUARE

1. Hypothesis is rejected. Sample report follows. Student reports will vary.

Forty-two undergraduate students were questioned concerning their plans for graduate education within the accounting field. Of these students, 16 were enrolled in the traditional day program and 26 were enrolled in the career evening program of study. Plans of study were coded as within 1 year of graduation, within 5 years of graduation, or no plans for enrollment in a graduation education program in accounting. In total, 32 students indicated that they planned to enroll in a graduate program in accounting within one year of graduation, see Table 6A. The results of the chi-square statistic indicate that there were no differences in student plans for graduate education by current study program ($\chi^2 = 1.3125$, $df = 2$, $p = 0.5188$).

3. Hypothesis is accepted. Student reports will vary.

4. Hypothesis is accepted. Sample report follows. Student reports will vary.

A voluntary Sexual Harassment Awareness Lecture was offered to the 39 employees working at site #236. Of these 39 employees, 34 (87%) elected to attend the lecture. All 25 male employees and 9 of the 14 female employees elected to attend the lecture, see Table 6D. A chi-square statistic was computed to determine if employee gender was related to lecture attendance. The results of the chi-square indicate that lecture attendance was significantly related to employee gender ($\chi^2 = 10.24$, $df = 1$, $p = 0.00137$). Significantly more female employees elected not to attend the lecture than were expected.

CHAPTER 7 – *t* TESTS OF TWO MEANS

1.
    a. paired, one-tailed
    b. independent, two-tailed
    c. independent, one-tailed
    d. independent, two-tailed

2. Hypothesis is accepted. Student reports will vary.

3. Hypothesis is accepted. Sample report follows. Student reports will vary.

Data was collected to determine if the length of the employee contract was related to union membership status. A total of 176 employee records were studied to determine the number of years of their contract and their union membership status. As shown in Table 7B, 90 employees were union members and 86 employees were not union members. Average contract length for union members was 3.8222 years ($SD = 1.232$). Average contract length for nonunion members was 4.1860 years ($SD = 1.101$). The results of the t test indicate that contract length for union members was significantly shorter than the contract length for nonunion members ($t = -2.06$; $p = 0.041$).

CHAPTER 8 – ANOVA ANALYSIS OF VARIANCE

1. Hypothesis is accepted. Sample report follows. Student reports will vary.

Combined waiting and treatment time for patients visiting the medical office was measured for a period of five days, Monday through Friday. Average wait/treatment time ranged from 52.18 minutes ($SD = 13.24$) on Tuesday, to 69.91 minutes ($SD = 14.72$) on Friday (see Table 8A1). The results of the ANOVA indicate that combined waiting and treatment time for patients seen on Friday was significantly greater than the combined waiting and treatment time for patients seen on Tuesday ($F = 3.6391$; $p = 0.0090$). Combined waiting and treatment times for the other days of the week did not vary significantly.

3. Hypothesis is rejected. Sample report follows. Student reports will vary.

Employee satisfaction was measured for employees working on the first ($n = 69$), second ($n = 36$), and third ($n = 30$) shift within the organization. Satisfaction measures for these three shifts ranged from a low of 66.81 for the second shift, to a high of 71 for the third shift. Average employee satisfaction for the three shifts was compared using ANOVA. As shown in Table 8C1, the results of the ANOVA indicate that employee satisfaction does not significantly differ by shift ($F = 0.8815$; $p = 0.4166$).

CHAPTER 9 – CORRELATION

1. Student hypotheses, reports, and tables will vary. One example follows.

Hypothesis 9A1 – There is a relationship between employee satisfaction and continuance commitment.

Two hundred and twenty-five employees were surveyed concerning their level of satisfaction and continuance commitment. Average employee satisfaction was reported as 69.609 points ($SD = 12.596$) and average continuance commitment was reported as 34.363 points ($SD = 8.908$). A correlation was calculated to determine if there was a significant relationship between employee satisfaction and continuance commitment. The results of the correlation ($r = -0.0625$; $p = 0.1810$) indicate that there is not a significant relationship between employee satisfaction and continuance commitment, see Table 9A1.

Table 9A1
Employee Attitudes
Correlation and Descriptive Statistics

	Mean	SD	Correlation Satisfaction
Satisfaction	69.609	12.596	
Continuance Commitment	34.363	8.908	-0.0625*

$n = 225$; *$p = 0.1810$

2. Acceptance or rejection is standard. Student reports and tables will vary. One example follows.

   a. Hypothesis is accepted.

The sample included 66 patients completing an in-patient substance abuse program. Their average age was 32.525 years ($SD = 7.637$). Patients were surveyed concerning their intentions to remain clean following discharge ($\bar{x} = 8.545$; $SD = 5.106$).

A correlation was calculated to determine if there is a significant relationship between patient age and intention to remain clean following discharge. The results of the correlation indicate that older patients were significantly less likely to express an intention to remain clean than were younger patients ($r = -0.3286$; $p = 0.005$), see Table 9B1.

Table 9B1
Substance Abuse Admissions
Correlation and Descriptive Statistics

	Mean	SD	Correlation Age
Age	32.525	7.637	
Intention to Remain Clean	8.545	5.106	-0.3286*

$n = 66$; *$p = 0.005$

   b. Hypothesis is rejected.
   c. Hypothesis is rejected.

CHAPTER 10 – TESTING YOUR SCALES

3. Sample report follows. Student reports will vary, recommendations should be standard.

A sample of 77 subjects completed the test-retest survey on Intrapersonal Role Conflict. The results of the test-retest correlation ($r = 0.8085$) indicate sufficient test-retest reliability for the scale to be used in hypothesis testing, see Table 10A.

4. Sample report follows. Student reports will vary, recommendations should be standard.

The internal consistency of the Continuance Commitment scale was tested with a sample of 61 subjects. Using the eight items, the Cronbach's alpha was calculated as 0.6875. The results indicate that of the eight items, one item (number 4) did not fit in well with the other items. Proper recoding standards should be verified for item 4. If recoding standards have been met, it is recommended that item 4 be removed and the internal reliability be recalculated with the remaining seven items. It is expected that the shorter scale will meet minimum expectations of internal consistency for use in hypothesis testing, see Table 10B.

Glossary

ANOVA, one-way – One-way Analysis of Variance. A statistical test that tests hypotheses which have one categorical variable which divides the sample into three or more groups and one continuous variable.

Attitude scales – a series of questions which, when summed, measure a variable describing a person's attitude. They tend to be classified as continuous variables.

Bivariate hypothesis – a testable statement which specifies the relationship between two variables.

Categorical variable – a variable which divides the values of the variable into independent groups (i.e. gender).

Cell – one box or number placement on a chi-square or cross-tabulation table.

Central tendency – the tendency for a group of numbers to fall towards the center of the distribution. Commonly measured by mean, median, and mode.

Chi-square – (χ^2) a univariate, bivariate, or multivariate (3 variable limit) statistic that tests hypotheses which have categorical variables (pronounced ki as in kite). Tests whether what was observed is different from what was expected.

Classification of your variables – the labeling of each variable as either categorical or continuous so that proper statistical techniques can be selected.

Coding – the assigning of numbers to variable values.

Computerized statistical packages – software that performs basic or advanced statistical computations using data supplied by the user.

Content validity – the appearance that the scale measures what it was designed to measure. Also called face validity.

Continuous variable – a variable in which levels of the variable exist between the values specified (i.e. age).

Correlation – (r) a statistic which measures the linear relationship between two continuous variables. Used in hypothesis testing.

Cross tabulation – a specialized frequency distribution which simultaneously displays the frequency of two or three variables in relation to each other.

Data entry – the process of inputting the raw data collected into the software program.

Data set – the entire collection of information gathered for the study.

Degrees of freedom – (df) a statistical measure calculated, in part, by the sample size or number of groups measured. Needed to determine the probability level in hypothesis testing.

Descriptive statistics – statistics which describe your data set. Often includes measures of central tendency and dispersion.

Dispersion – a description of how far the data is scattered along the values collected. Commonly measured by range, standard deviation, and variance.

Face validity – the appearance that the scale measures what it was designed to measure. Also called content validity.

Frequency distribution – an arrangement, in table form, of the raw data collected. Often provides counts and percentages.

Internal reliability – internal consistency from one item within the scale to another. A measurement of reliability. Often measured using Cronbach's coefficient alpha. Should be calculated for every scale.

Mean – ($\bar{x}$) arithmetic average. A measure of central tendency.

Median – the middle number of an ordered distribution. A measure of central tendency.

Missing values – variables or questions which were not answered or collected for every item or person in your sample.

Mode – the value most frequently occurring in a distribution. A measure of central tendency.

Multivariate hypothesis – a hypothesis which specifies the relationship between three or more variables.

Negative Correlation – a statistical relationship between two continuous variables which indicates that they move in opposite directions. When one of the variables increases, the other variable decreases. A perfect negative correlation is -1.00.

One-tailed test – a statistical test for a hypothesis written with a specific direction of differences proposed.

One-way Analysis of Variance – a statistic that tests hypotheses which have one categorical variable which divides the sample into three or more groups and one continuous variable.

Paired test – a statistic (typically a t test) which compares the subjects' scores from two different times. Often used in the test-retest design.

Positive Correlation – a statistical relationship between two continuous variables which indicates that they both move in the same direction. When one of the variables increases, the other variable also increases. A perfect positive correlation is 1.00.

Probability – (p) a number between 0.00 and 1.00 which indicates the likelihood that no differences exist in the data you have tested. Also called the significance level.

Psychometric properties – the results of psychometric testing of a scale.

Psychometric testing – testing of measurement scales to demonstrate reliability and validity.

Random assignment – the equal, unbiased chance of being assigned to a particular group.

Random selection – the equal, unbiased chance of being selected.

Range – the difference between the largest and smallest value observed for a variable. A measure of dispersion.

Raw data – the information collected for testing, may be on a survey or observation tally sheet.

Reliability – the extent that the measure gives similar answers with repeated uses.

Respondents – people who participate in the study. Also called subjects or participants.

Reverse score – to recode values such that higher scores are given lower values and lower scores are given higher values (i.e. 5 = 1, 4 = 2, 3 = 3, 2 = 4, 1 = 5).

Sample size – (n) the number of items in the group or people in your sample for which you have collected data.

Scales – variables which are measured by a series of related questions which, when summed or averaged, provide a single measure.

Sensitivity – the ability of the scale to measure small changes when changes actually take place in the item being measured.

Significance level – (p) a number between 0.00 and 1.00 which indicates the likelihood that no differences exist in the data you have tested. Also called the probability.

Standard deviation – (SD) the square root of the variance. A measure of dispersion.

Statistical tests – a series of tests which produce mathematical results to aid in decision making when analyzing data.

Summary statistics – statistics which summarize the data for ease of study. May include frequencies and percentages.

Test-retest reliability – the extent that the scale gives the same answers with repeated uses. Important to demonstrate during scale development.

t Test – a bivariate statistical test for a hypothesis which has one categorical variable with two values and one continuous variable.

Two-tailed test – a statistical test for a hypothesis written without a specific direction of differences proposed.

Validated measurements – measurement scales which have been demonstrated to be both reliable and valid.

Validity – the accuracy of the scale to measure what is was designed to measure. Important to demonstrate during scale development.

Variables – measurements of characteristics, qualities, or ideas used in hypothesis testing.

Variance – The average of squared differences each value is from the mean. A measure of dispersion.

INDEX

A

analysis of variance, 45, 47, 50
ANOVA (see analysis of variance), 7, 23, 45, 46, 47, 48, 49, 50, 75, 76, 77, 78, 83, 88, 91

B

bell curve (see also normal distribution), 26
bivariate hypothesis, 1, 8

C

categorical variable, 5, 6, 7, 8, 9, 18, 20, 25, 29, 31, 37, 39, 40, 45, 47, 48, 70, 85, 91, 93, 94
central tendency, 17, 18, 19, 20, 92
chi-square, 7, 13, 23, 29, 30, 31, 33, 34, 35, 83, 87, 91
classifying variable, 31
coding, 12, 25, 62, 70, 86
confidence interval, 46, 47, 48, 49
content validity, 62, 92
continuous variable, 5, 6, 7, 8, 9, 11, 18, 19, 20, 25, 37, 39, 40, 45, 47, 48, 52, 56, 91, 92, 93, 94
correlation, 7, 9, 23, 51, 52, 53, 54, 56, 57, 58, 59, 60, 63, 78, 79, 83, 88, 89, 90
cross tabulation, 13, 14, 16, 35
Cronbach's coefficient alpha, 90
cumulative values, 20

D

degrees of freedom, 24, 29, 31, 37, 45
descriptive statistics, 13, 17, 19, 20, 54, 68, 69
dispersion, 17, 19, 20, 92, 93, 94

E

expected values, 32

F

face validity, 62, 91
frequency distribution, 11, 12, 13, 16, 17, 18, 19, 20, 27, 68, 92

I

internal reliability, 60, 61, 62, 80, 90

M

mean, 17, 18, 19, 37, 40, 45, 46, 47, 51, 52, 62, 91, 94
median, 17, 18, 19, 20, 91
missing values, 24
mode, 17, 18, 19, 20, 85, 91
multivariate hypothesis, 8, 45

N

negative correlation, 51, 56, 93
normal distribution (see also bell curve), 26
null hypothesis, 23

O

observed values, 32
one-tailed probability, 42, 44
one-tailed test, 26, 41, 42, 43, 44
one-way analysis of variance, 45, 47

P

paired test, 37
positive correlation, 51, 53, 93
probability (see also significance level), 23, 24, 27, 29, 30, 31, 32, 34, 35, 37, 38, 39, 41, 42, 44, 45, 46, 48, 49, 50, 51, 52, 53, 54, 57, 58, 70, 92, 94
psychometric properties, 62, 63
psychometric testing, 60, 93

Q

quality control, 11

R

random assignment, 43
range, 6, 12, 13, 17, 19, 27, 34, 47, 92
raw data, 38, 92
recoding (see also reverse score), 27, 71, 90
reliability, 59, 60, 61, 62, 63, 80, 90, 92, 93, 94
respondents, 63, 68, 84

reverse score (see also recoding), 25, 86

S

sample size, 20, 21, 24, 46, 51, 53, 54, 57, 59, 92
scatter plot, 52, 56
sensitivity, 59
significance level (see also probability), 23, 29, 31, 35, 37, 44, 45, 50, 51, 58, 93
standard deviation, 17, 20, 21, 37, 45, 54, 92
summary statistics, 12

T

t test, 7, 8, 9, 23, 37, 38, 39, 40, 41, 42, 43, 44, 45, 74, 83, 88, 93
test-retest reliability, 59, 60, 63, 80, 90
two-tailed test, 26, 43, 44, 71, 74

V

variance, 17, 19, 20, 92, 94